Advance in Recovery and Application of Bioactive Compounds from Seafood

Advance in Recovery and Application of Bioactive Compounds from Seafood

Editors

Charlotte Jacobsen
Susan Løvstad Holdt

MDPI • Basel • Beijing • Wuhan • Barcelona • Belgrade • Manchester • Tokyo • Cluj • Tianjin

Editors

Charlotte Jacobsen
Technical University of Denmark
Denmark

Susan Løvstad Holdt
Technical University of Denmark
Denmark

Editorial Office
MDPI
St. Alban-Anlage 66
4052 Basel, Switzerland

This is a reprint of articles from the Special Issue published online in the open access journal *Foods* (ISSN 2304-8158) (available at: https://www.mdpi.com/journal/foods/special_issues/Recovery_Application_Bioactive_Compounds_Seafood).

For citation purposes, cite each article independently as indicated on the article page online and as indicated below:

LastName, A.A.; LastName, B.B.; LastName, C.C. Article Title. *Journal Name* **Year**, *Volume Number*, Page Range.

ISBN 978-3-0365-4587-5 (Hbk)
ISBN 978-3-0365-4588-2 (PDF)

Contents

About the Editors

Charlotte Jacobsen

Charlotte Jacobsen, Professor: She leads the Research group for Bioactives—Analysis and Applications at the National Food Institute, Technical University of Denmark. She is internationally renowned for her research in lipid oxidation of omega-3 rich foods and she has received several awards, including the Danish Danisco price in 2003 (USD 40.000), the French La Médaille Chevreul in 2010 awarded by the Association Francaise pour l'étude des Corps Gras, the German DGF Normann Medaille in 2020, the Stephen S. Chang award (2021) and AOCS Fellow (2022), and two best paper awards from the American Oil Chemist's society. She was appointed by EFSA as an expert in the Fish Oil working group under the Biohazard Panel to evaluate the potential hazards associated with the human intake of refined fish oil. Her recent research places focus on the use of side-streams from plant and marine food production as new ingredients (e.g., antioxidants) for the food industry. She has led several large national and international projects, including the on-going EU BBI JU project WaSeaBi. Most of her projects involve collaboration with industry partners, e.g., Royal Greenland, Kalsec, Nestlé and Novozymes. Her publication list includes more than 250 peer-reviewed manuscripts and book chapters.

Susan Løvstad Holdt

Susan Løvstad Holdt, Associate professor in the Research Group Bioactives—Analysis and Application at The National Food Institute, Technical University of Denmark (DTU Food). SLH is an expert in research regarding seaweed composition, bioactives and technologies for extraction. Furthermore, she is involved in research on seaweed as a food, legislation, standardization, retaining the qualities of seaweed, but also with a focus on reducing the contaminants found in seaweed. SLH has led or taken part in research projects involving companies such as CP Kelco and Orkla, and she is Secretary General of the International Seaweed Association and chair of the Seaweed Network in Denmark. SLH's list of publications include FAO reports and book chapters, and has more than 30 peer-reviewed publications counting the highly cited (>900) review on 'Bioactive compounds in seaweed: functional food applications and legislation' (Holdt and Kraan, 2011).

foods

MDPI

Editorial

Introduction to the Special Issue: "Advance in Recovery and Application of Bioactive Compounds from Seafood"

Charlotte Jacobsen * and Susan L. Holdt

Research Group for Bioactives–Analysis and Analysis, National Food Institute, Technical University of Denmark, 2800 Kgs. Lyngby, Denmark; suho@food.dtu.dk

* Correspondence: chja@food.dtu.dk; Tel.: +45-23279075

Due to increased focus on a circular bioeconomy, full utilization of marine biomass, including side streams from the seafood processing industry as well as utilization of biomass that has not been used to a great extent in the Western world (e.g., seaweed), is receiving increased attention from both academia and industry. These forms of marine biomass contain a wide range of interesting biomolecules with beneficial health and/or functional properties, which can be exploited for applications in food, feed, dietary supplements or pharma. However, there are several challenges to be overcome. Six high quality research papers about opportunities and challenges in this area of research have been published in this Special Issue.

Naseri and colleagues developed a multi-extraction process in which protein was extracted from the commercial red seaweed *Euchema denticulatum*, followed by the extraction of carrageenan which is widely used in, for example, the food industry [1]. The process included enzyme assisted aqueous extraction followed by alkaline extraction using N-acetyl-cysteine. Up to almost 60% of the protein in the seaweed could be recovered depending on the enzyme used.

Seaweed also contains minerals such as iodine. However, in some species the iodine content can be so high that it becomes a concern for the consumer. Nielsen and colleagues studied how the content of iodine can be reduced in the brown seaweed *Saccharina latissima* (sugar kelp) by water blanching or freezing to avoid an excessive intake of iodine via the consumption of this seaweed [2]. They found that the highest reduction in iodine content could be obtained by water blanching at 80 °C for 120 s. They also studied whether water blanching affected the content of other valuable compounds in the seaweed.

It is well known that fish oil is healthy due to its high content of omega-3 polyunsaturated fatty acids. However, the polyunsaturated nature of fish oil also makes it highly susceptible to lipid oxidation, which leads to formation of undesirable off-flavors. This limits the use of fish oil for food applications. Therefore, strategies to reduce lipid oxidation are necessary. One possible strategy is to emulsify the fish oil in a delivery system which protects the oil against lipid oxidation. This approach requires emulsifiers with good emulsifying properties for preparing emulsions with high physical and oxidative stability. Padial-Dominguez and co-workers optimized the enzymatic hydrolysis conditions used for the preparation of hydrolysates of whey protein and soy protein with optimal emulsifying properties [3]. They also investigated the physical and oxidative stability of fish oil-in-water emulsions prepared with soy protein hydrolysates. They found that soy protein hydrolysates reduced lipid oxidation in such emulsions compared to the use of non-hydrolyzed soy protein isolate. Rahmano-Manglano and colleagues investigated another approach to stabilize fish oil against oxidation, namely, microencapsulation [4]. They studied the influence of the carbohydrate-based wall matrix (glucose syrup or maltodextrin) and the storage temperature (4 or 25 °C) on the oxidative stability of microcapsules, prepared by spray drying emulsions with whey protein hydrolysates as the emulsifier. They found that glucose syrup provided the highest oxidative stability.

Citation: Jacobsen, C.; Holdt, S.L. Introduction to the Special Issue: "Advance in Recovery and Application of Bioactive Compounds from Seafood". *Foods* **2021**, *10*, 266. https://doi.org/10.3390/foods10020266

Received: 11 January 2021
Accepted: 20 January 2021
Published: 28 January 2021

Apart from having emulsifying properties, some protein hydrolysates also have antioxidative properties. Tabakaeva and colleagues studied the antiradical properties of hydrolysates and hydrothermal extracts of bivalve mollusks (*Anadara broughtonii*). They also evaluated their effect on lipid oxidation and lipolysis in mayonnaise [5]. They studied hydrolysates obtained either by acid or enzymatic hydrolysis and observed that acid hydrolysates were more efficient than enzymatic hydrolysates in preventing oxidation and lipolysis in mayonnaise.

Protein hydrolysates may also have bioactive properties, and this was studied in the last paper in this Special Issue. Kang, Skanberg and Myracle studied the anti-hyperglycemic effects including α-glucosidase, α-amylase, and dipeptidyl peptidase-IV (DPP-IV) inhibitory activities and glucagon-like 1 (GLP-1) secretory activity of green crab hydrolysates obtained with different commercially available enzymes [6]. They concluded that hydrolysates obtained by Protamex treatment have the potential for application in type 2 diabetes management.

We hope that this Special Issue will further promote the interest and research in bioactive compounds from seafood. This Special Issue has only scratched the surface, and there is much more to investigate about the possible uses of marine bioactive compounds for feed, food, cosmetics, and pharma; thereby, bringing us closer to an optimized system for the utilization of our marine resources, in order to achieve a sustainable future based on a circular bioeconomy.

Conflicts of Interest: The authors declare no conflict of interest.

References

1. Naseri, A.; Jacobsen, C.; Sejberg, J.J.P.; Pedersen, T.E.; Larsen, J.; Hansen, K.M.; Holdt, S.L. Multi-Extraction and Quality of Protein and Carrageenan from Commercial Spinosum (*Eucheuma denticulatum*). *Foods* **2020**, *9*, 1072. [CrossRef] [PubMed]
2. Nielsen, C.W.; Holdt, S.L.; Sloth, J.J.; Marinho, G.S.; Sæther, M.; Funderud, J.; Rustad, T. Reducing the High Iodine Content of *Saccharina latissima* and Improving the Profile of Other Valuable Compounds by Water Blanching. *Foods* **2020**, *9*, 569. [CrossRef] [PubMed]
3. Padial-Domínguez, M.; Espejo-Carpio, F.J.; Pérez-Gálvez, R.; Guadix, A.; Guadix, E.M. Optimization of the Emulsifying Properties of Food Protein Hydrolysates for the Production of Fish Oil-in-Water Emulsions. *Foods* **2020**, *9*, 636. [CrossRef] [PubMed]
4. Rahmani-Manglano, N.E.; González-Sánchez, I.; García-Moreno, P.J.; Espejo-Carpio, F.J.; Jacobsen, C.; Guadix, E.M. Development of Fish Oil-Loaded Microcapsules Containing Whey Protein Hydrolysate as Film-Forming Material for Fortification of Low-Fat Mayonnaise. *Foods* **2020**, *9*, 545. [CrossRef] [PubMed]
5. Tabakaeva, O.V.; Piekoszewski, W.; Kalenik, T.K.; Maximova, S.N.; Tabakaev, A.V.; Poleshyk, D.V.; Proniewicz, L. Antiradical Activity of Hydrolysates and Extracts from Mollusk *A. broughtonii* and Practical Application to the Stabilization of Lipids. *Foods* **2020**, *9*, 304. [CrossRef] [PubMed]
6. Kang, B.; Skonberg, D.I.; Myracle, A.D. Anti-Hyperglycemic Effects of Green Crab Hydrolysates Derived by Commercially Available Enzymes. *Foods* **2020**, *9*, 258. [CrossRef] [PubMed]

foods

MDPI

Article

Multi-Extraction and Quality of Protein and Carrageenan from Commercial Spinosum (*Eucheuma denticulatum*)

Alireza Naseri [1], Charlotte Jacobsen [1,*], Jimmy J. P. Sejberg [2], Tommy Ewi Pedersen [2], Jan Larsen [2], Karin Meyer Hansen [2] and Susan L. Holdt [1]

[1] Research Group for Bioactives—Analysis and Application, The National Food Institute, Technical University of Denmark (DTU Food), Kemitorvet, Building 204, DK-2800 Kgs. Lyngby, Denmark; alireza@food.ku.dk (A.N.); suho@food.dtu.dk (S.L.H.)

[2] Innovation, CP Kelco, Ved Banen 16, DK-4623 Lille Skensved, Denmark; jimmy.sejberg@cpkelco.com (J.J.P.S); tommy.ewi.pedersen@gmail.com (T.E.P.); jan.larsen@cpkelco.com (J.L.); KarinMeyer.hansen@cpkelco.com (K.M.H.)

* Correspondence: chja@food.dtu.dk

Received: 16 June 2020; Accepted: 3 August 2020; Published: 6 August 2020

Abstract: Seaweeds contain many valuable compounds that can be used in the food industry. Carrageenan is a polysaccharide which has been extracted from seaweed for centuries and is used as a texturizer in food and non-food products. However, seaweeds contain compounds other than carrageenan, such as proteins, which could also be extracted. This extraction should be done without compromising the industrial scale carrageenan extraction yield and quality. This study aimed at up-stream protein extraction from red seaweed *Eucheuma denticulatum* by using of an optimized enzyme-assisted extraction, including of an aqueous/enzymatic treatment followed by alkaline extraction, and then the commercial carrageenan extraction. The protein extraction efficiency of four enzymes was evaluated including Celluclast® 1.5L, Shearzyme® 500 L, Alcalase® 2.4 L FG and Viscozyme® L at a concentration of 0.0, 0.1, 0.2 and 0.4% (*w*/*w*). To avoid detrimental effects on carrageenan, all the experiments were performed at pH 7 at room temperature. The results showed that 0.2% *w*/*w* Alcalase® or Viscozyme® added individually achieved the highest protein extraction efficiencies (59 and 48%, respectively) at pH 7 and room temperature ($p < 0.05$). Determination of the most common carrageenan quality parameters indicated that using any of these enzymes had no negative effect on the carrageenan yield and quality.

Keywords: algae; industrial seaweeds; vegan protein; combined extraction; bioactive compounds; sustainability; bioeconomy; functionality

1. Introduction

Seaweeds contain many industrial ingredients, but also other compounds of interest, including bioactive compounds, which can be used in the food industry. Carrageenan is a polysaccharide, which has been extracted from seaweed for centuries and is used as a texturizer in food and non-food products. *Eucheuma denticulatum*, with the commercial name spinosum, is one of the main red seaweeds used in carrageenan production, and this seaweed constituted almost 20% of the carrageenan-containing seaweeds harvested worldwide in 2015 [1]. Methods of extracting carrageenan from seaweed are well known, and such methods involve alkaline extraction of seaweed at high temperatures (generally around 100–120 °C) [2]. However, the commercially used alkaline extraction of seaweed usually results in a carrageenan product, which also contains other compounds, such as proteins, antioxidants, and pigments [3].

It would be advantageous to have a method of preparing carrageenan, where other compounds including bioactive compounds from seaweeds could be extracted as value-added products, without decreasing the carrageenan yield and the functional quality of the carrageenan.

Furthermore, the global demand for protein is increasing and is expected to escalate further in the coming decades—mostly due to population growth, which must be matched by increased food production [4]. There is a good reason to develop new technologies for the industrial extraction of vegan proteins from seaweeds, since some species of seaweed have an interesting amino acid profile close to that of animal protein. For example, the ratio between essential amino acids (EAA) and total amino acids (AA) in *Palmaria palmata* was 44–53% while this score for soya and egg protein was 39 and 47% respectively [5]. The global carrageenan production in 2014 amounted to 60,000 tonnes, with a value of US$ 626 million in 2014. From these numbers, it can be estimated that the total dried seaweed consumption for this production, was at least 300,000 tonnes per year. The protein content of these types of seaweed are in the range of 4–28% [6]. If just half of the total amount of protein could be extracted, more than 20,000 tonnes of a new high-value protein product would be obtained per year. [6,7] The variation in protein content in seaweeds used in carrageenan production could be attributed to differences in seasonality, growth conditions in the environment, or source and species of resource [6,8].

Although *E. denticulatum* with 4–5% protein content has the lowest level of protein in comparison with other seaweeds such as *Palmaria palmata*, *Furcellaria lumbricalis* or *Chondrus crispus*, it should be considered that the global industrial utilization of this seaweed for carrageenan production was 45000 dry tonnes in 2015 which is equal to 20% of total harvest. Therefore, large amounts of industrial seaweed are available [1,6]. Hence, an improved method of extracting carrageenan from seaweed would be advantageous, and in particular, a method enabling the extraction of proteins and other bioactive compounds in addition to carrageenan, without any detrimental effects on carrageenan yield and quality.

In recent years, varieties of processes for the extraction of protein from various seaweed species have been reported. These processes include enzyme-assisted extraction (EAE) [7,9,10], physical processes [11,12], chemical extraction [13,14], as well as novel techniques such as ultrasound-assisted extraction (UAE) [15,16], pulsed electric field (PEF) extraction [12,17], and microwave-assisted extraction (MAE) [4,18].

Some of these studies evaluated the effect of enzymatic treatment, alkaline pretreatment, and process conditions on viscosity and gel strength of carrageenan; extracted from different red seaweeds commonly used in carrageenan production. As an example, Azevedo et al. (2015) studied the effect of pre-extraction alkali treatment on the chemical structure and gelling properties of extracted hybrid carrageenan from *C. crispus* and *Ahnfeltiopsis devoniensis* [19]. The results showed that increasing the KOH content, and the pre-treatment time improved the gelling properties in both seaweeds [19]. In addition, the effect of process conditions on the viscosity and gel strength of semi-refined carrageenan (SRC), produced from red seaweed *Kappaphycus alvarezii* was studied by Anisuzzaman et al. in 2014. The experimental results showed that gel viscosity increased with the decrease of cooking time, cooking temperature and potassium hydroxide (KOH) concentration (% *w*/*w*). In contrast, gel strength increased when cooking time, cooking temperature and KOH concentration (% *w*/*w*) increased. [20]. Moreover, in another study, the development of a high yielding carrageenan extraction method from *Eucheuma cottonii* using cellulase and the fungi *Aspergillus niger* was investigated [21]. However, no other studies have dealt with both protein and carrageenan extraction from *E. denticulatum*, as well as the effects of the protein extraction on the yield and quality of the extracted carrageenan.

The overall aim of this study was to design and develop a method to extract more than one product (multi-extraction), e.g., extraction of both protein and carrageenan, from *E. denticulatum* (spinosum). The extraction of protein from spinosum upstream the carrageenan extraction was optimized with regard to temperature, pH, the concentration of enzymes, and extraction time. The selected enzymes were in accordance with the previous study, which was successfully carried out on *P.palmata* [5]. The quality of the extracted protein was evaluated based on the amino acid profile, and the contents of essential amino acids (EAA). In addition, the effect of multi-extraction on the yield and quality of the isolated carrageenan (i.e., gel strength, turbidity, and viscosity) was evaluated.

2. Material and Methods

2.1. Seaweed Biomass and Preparation

CP Kelco (Lille Skensved, Denmark) provided the *Eucheuma denticulatum* seaweed used in their carrageenan production line. The seaweed biomass was harvested and shipped in semi-dried condition in big pack pallets from the Philippines and Indonesia to CP Kelco. The pallets were stored in a non-insulated warehouse at ambient outdoor temperature. The industrial samples were taken from different batches to obtain representative samples. All seaweed samples were dried at ambient temperature (as is the procedure in industrial-scale carrageenan production), after which the biomass was reduced to 0.5–1 cm particle size using a cutting mill (SM 2000, Retsch, Haan, Germany) in the lab of DTU Food. The milled seaweed was kept in plastic bags in a freezer at −20 °C until used. Dry matter (DM) content for all samples in this study was determined by drying the biomass in an oven at 105 °C until constant weight.

2.2. Enzymes and Chemicals

All enzymes, including a cellulase Celluclast® 1.5 L, the xylanase Shearzyme® 500 L, the protease (endo-peptidase) Alcalase® 2.4 L FG, and the multi-enzyme mixture Viscozyme® L which contains a wide range of carbohydrases, including arabinase, cellulase, β-glucanase, hemicellulase, and xylanase, were provided by Novozymes A/S (Bagsværd, Denmark). All chemicals used in this study were from Merck (St. Louis, IL, USA). All solvents used were HPLC grade and purchased from Lab-Scan (Dublin, Ireland). The standards for amino acids analysis were purchased from Sigma-Aldrich (St. Louis, MO, USA). HPLC grade water was prepared by a Milli-Q® Advantage A10 water deionizing system from Millipore Corporation (Billerica, MA, USA).

2.3. Aqueous/Enzymatic Extraction

Semi-dried seaweed samples (25 g) were weighed in 1 L Erlenmeyer flasks. In accordance with the selected ratio in Table 1, deionized water was added to each of 1 L Erlenmeyer flasks. The sample was rehydrated for 1 h, and the enzymes were then added. Carrageenan releases from seaweed by increasing the temperature, and the enzymatic treatment was therefore performed at room temperature (20–22 °C), in order not to obtain simultaneous carrageenan and protein extraction. All treatments were run in triplicates (n = 3).

Table 1. Protein extraction conditions (ratio, extraction time and enzymes concentration (on weight basis (*w*/*w*%)) and the average extraction efficiency (%) ± standard deviations of treated *Eucheuma denticulatum* seaweed. Different superscripts letters in the same column indicate significant difference at the confidence level of 95% (α = 0.05; n = 3).

Number	Ratio (Seaweed: Water)	Extraction Time (h)	The Concentration of Celluclast® Plus Shearzyme® (*w*/*w*%)	Extraction Efficiency (%)
1	1:15	4	0	7.2 ± 0.04 [d]
2	1:15	4	0.1	10.9 ± 0.28 [d]
3	1:15	4	0.2	9.2 ± 0.29 [d]
4	1:15	4	0.4	12.7 ± 0.42 [d]
5	1:15	6	0	8.1 ± 0.78 [d]
6	1:15	6	0.1	9.0 ± 0.32 [d]
7	1:15	6	0.2	12.3 ± 0.09 [d]
8	1:15	6	0.4	11.9 ± 0.29 [d]
9	1:15	8	0	8.8 ± 0.28 [d]
10	1:15	8	0.1	9.4 ± 0.32 [d]
11	1:15	8	0.2	14.5 ± 0.43 [a,b,c]
12	1:15	8	0.4	13.7 ± 0.42 [d]

Table 1. *Cont.*

Number	Ratio (Seaweed: Water)	Extraction Time (h)	The Concentration of Celluclast® Plus Shearzyme® (w/w%)	Extraction Efficiency (%)
13	1:20	4	0	7.4 ± 0.19 [d]
14	1:20	4	0.1	7.3 ± 0.18 [d]
15	1:20	4	0.2	9.9 ± 0.28 [d]
16	1:20	4	0.4	12.7 ± 0.30 [d]
17	1:20	6	0	10.1 ± 0.44 [d]
18	1:20	6	0.1	11.5 ± 0.31 [d]
19	1:20	6	0.2	15.1 ± 0.52 [a,b]
20	1:20	6	0.4	15.5 ± 0.11 [a]
21	1:20	8	0	8.8 ± 0.41 [d]
22	1:20	8	0.1	11.6 ± 0.12 [d]
23	1:20	8	0.2	12.9 ± 0.11 [d]
24	1:20	8	0.4	14.2 ± 0.04 [b,c]
25	1:25	4	0	8.9 ± 0.49 [d]
26	1:25	4	0.1	8.9 ± 1.03 [d]
27	1:25	4	0.2	12.9 ± 0.25 [d]
28	1:25	4	0.4	11.8 ± 2.03 [d]
29	1:25	6	0	9.3 ± 0.55 [d]
30	1:25	6	0.1	9.8 ± 0.64 [d]
31	1:25	6	0.2	10.7 ± 0.01 [d]
32	1:25	6	0.4	12.9 ± 0.43 [d]
33	1:25	8	0	9.4 ± 0.59 [d]
34	1:25	8	0.1	10.7 ± 1.08 [d]
35	1:25	8	0.2	12.1 ± 0.22 [d]
36	1:25	8	0.4	12.6 ± 0.73 [d]

Different letters indicate significant differences ($p < 0.05$).

2.4. NAC-Assisted Alkaline Extraction

The conditions for the alkaline extraction were modified from those reported by Harnedy and FitzGerald [22] and were done in accordance with the study by Naseri et al. for the extraction of protein from *P. palmata* [5]. In summary, after the enzymatic treatment (see Section 2.3), the alkaline extraction was don and repeated three times with a solution containing 1 g/L of NAC and 4 g/L of NaOH. A laboratory orbital shaker was used at a speed of 120–130 rpm and ambient temperature for 1.5 h. (Figure 1). The liquid fractions recovered were pooled together with the liquid fraction recovered from the enzymatic extraction. The pooled liquid fraction was stored at 4 °C overnight before precipitation of protein by the addition of acid (HCl).

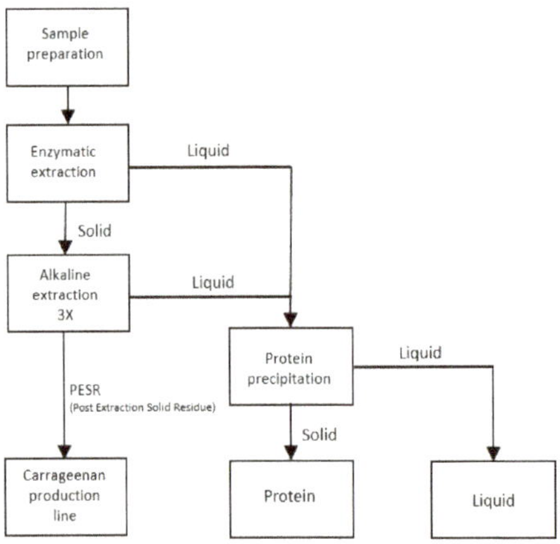

Figure 1. Process flow diagram for a proposed multi-extraction process of seaweed with both enzymatic extraction of proteins, followed by NAC assisted alkaline extraction before the actual industrial utilization of the spinosum seaweed for carrageenan production.

2.5. Post-Extraction Solid Residue (PESR)

At the end of the extraction process, the filter-cakes mentioned above were placed in plastic containers. In order to avoid negative effects of high temperature on protein quality, all filter-cakes were dried in an oven at 40–45 °C. Then, the samples were ground to powder and kept in the freezer at −20 °C prior to further analysis.

2.6. Experimental Plan

The proposed multi-extraction of proteins and carrageenan from spinosum seaweed is shown in Figure 1. This includes enzymatic treatment, NAC (see explanation above) assisted extraction and protein separation prior to the normal single product carrageenan production/extraction. In order to study and optimize this multi-extraction, the experimental setup consisted of five steps. The first step was to determine the best conditions for temperature, pH, the ratio of seaweed to water and enzyme combination. In accordance with the results reported in the literature [5,23], the ratios of seaweed to water selected were 1:15, 1:20 and 1:25. Enzyme concentrations of 0.0, 0.1, 0.2 and 0.4% *w*/*w* were used. All the enzyme treatments were carried out in duplicates for 4, 6 or 8 h at pH 7 to avoid detrimental effect on carrageenan quality.

In the second step, a variety of enzymes were evaluated at different concentrations followed by N-acetyl-L-cysteine (NAC) assisted extraction. NAC, L-cysteine-hydrochloride monohydrate and β-mercaptoethanol are reducing agents significantly increase the yield of alkaline soluble nitrogen [22]. Therefore, the combination of NaOH and NAC as a food-grade reducing agent was selected to increase the extraction of proteins in the current study. In this step, different enzymes were used at the selected optimal condition (Table 1), after which the effect of protein extraction on the yield of carrageenan extraction (step 3) and carrageenan quality (step 4) was evaluated. In the last step, the amino acid profiles for the selected treatments were compared. All treatments were run in triplicates (n = 3).

2.7. Protein Precipitation and Centrifugation

Protein precipitation was performed by lowering the pH of the solution to the isoelectric point of the protein (3.5). The pH of the pooled protein extract was lowered to 3.5 by the addition of aqueous hydrochloric acid (2 M). Subsequently, the mixture was centrifuged at 4400 g for 15 min at 4 °C and then the supernatant and solid residue were separated and stored in the freezer at −20 °C. The solid fractions were freeze-dried and milled prior to the analysis (*cf.* above).

2.8. Total Nitrogen, Protein Content and Extraction Efficiency

Total nitrogen content was measured by a DUMAS nitrogen/protein analyzer using a fully automated rapid MAX-N exceed (Elementar Analysensysteme GmbH, Langenselbold, Germany). For this purpose, 150–250 mg of dried fractions and 3–4 mL of liquid fractions were fed into the analyser. Protein content was calculated by multiplying the nitrogen content by a conversion factor of 5 [24]. All treatments were run in triplicates (n = 3).

The protein extraction efficiency for every treatment was calculated based on the below equation [5]:

$$\text{Extraction efficiency }\% = \frac{\text{Protein content before extraction} - \text{Protein content after extraction}}{\text{Protein content before extraction}} \times 100 \quad (1)$$

To do the calculation faster, the below equation was used:

$$\text{Extraction efficiency }(\%) = 100\% - \text{Protein recovered in PESR }\% \quad (2)$$

2.9. Carrageenan Extraction and Yield

After protein extraction, the seaweed samples were freeze-dried, milled and passed through a 1 mm mesh sieve to obtain uniform particle size. The carrageenan was extracted using a water-extraction technique. In brief, the seaweeds were soaked in milli-Q water overnight (5% *w*/*v*). The pH of the suspension was adjusted (pH 7.5–8.5) by KOH and the carrageenan was extracted at 99 °C for 1.5 h in a water bath with shaking. The carrageenan was separated from the seaweed residue by filtration using filter aid and subsequently isolated by precipitation in isopropanol at the ratio of 1:3 (seaweed/alcohol). The samples were freeze-dried, and yields were determined by weighing [25]. All treatments were run in triplicates ($n = 3$).

2.10. Carrageenan Gel Strength

The gelling properties of the isolated carrageenan was evaluated, as it would be performed in industry, in a dessert formulation (milk jelly) prepared from whole milk and sugar. The milk jelly ingredients were heated at 80 °C for 30 min, then filled directly into crystallization dishes. Finally, the gel was allowed to cool in a 5 °C water bath for 3–4 h and the top layer of the gel was then carefully removed using a wire cheese slicer. The gel and breaking strengths were measured with a Stable Micro System (SMS) Texture Analyser-TX.XT2 (Godalming, UK) with a plunger: 0.5 inch diameter; plunger speed: 1 mm/s; distance: 30 mm. The gel strength was measured at 4, 8 and 12 mm penetration of the gel. The breaking strength was measured at the first peak on the curve. The gel and breaking strengths were calculated as the averages of three measurements.

2.11. Turbidity of a 1.0% Solution of Carrageenan

The turbidity was determined using HACH 2100 N Turbidimeter. Turbidity was measured using a 1.0% (*w*/*w*) aqueous solution of carrageenan obtained in 2.9 which was prepared by the addition of 1.00 g carrageenan to 100 mL deionized water.

2.12. Food Chemicals Codex (FCC) Viscosity

The viscosity of the 1.5% aqueous solution of carrageenan was measured at 75 °C when making FCC-viscosity on a product standardized with sugar, and the amount of weighed material must be corrected to 1.5% pure material. FCC viscosity was measured at CP Kelco using a LVF viscometer (Brookfield, Middleboro, MA, USA.) fitted with a UL-adapter. The viscosity was measured at 30 rpm for 30 s.

2.13. Amino Acid Composition

The amino acid (AA) composition was determined as described by Farvin et al. (2010). To 50 mg dry weight of the sample was added 6 M aqueous hydrochloric acid, and the mixture was heated in the oven at 105 °C overnight. Following filtration through a 0.2 μm filter, derivatization was carried out using the EZ:Faast kit from Phenomenex A/S (Allerød, Denmark). The amino acid composition was determined using LC-MS (Agilent 1100 Series, LC/MSD Trap, Santa Clara, CA, USA) with an EZ:faast 4u AAA-MS new column (250 × 3.0 mm, Phenomenex) as described by Farvin et al. This procedure does, however, not allow for the detection of tryptophan (Trp) and cysteine (Cys) as both amino acids decompose during the acid hydrolysis [26].

2.14. Statistical Analysis

Statgraphics Centurion 18 (Statistical Graphics Corp., Rockville, MD, USA) was used for data analysis. Data were expressed as mean ± standard deviation, corresponding to three experimental replicates ($n = 3$). Firstly, by ANOVA test, multiple sample comparison analysis was performed to identify significant differences between samples. Secondly, mean values were compared using Duncan's test. Differences between means were considered significant at $p < 0.05$.

3. Results and Discussion

3.1. The Optimization of Parameters for Enzymatic Treatment

In the present study, industrially utilized semi-dried *Eucheuma denticulatum* (spinosum) containing approximately 50% water and 3.8% DM protein was used as a raw material. Carrageenan is sensitive to low pH and will start to extract at high temperatures. Hence, to avoid detrimental effects on carrageenan quality, all experiments were performed at room temperature (<30 °C) at pH ≥7. Based on preliminary experiments, a combination of Celluclast® and Shearzyme® was selected.

Table 1 shows the investigated process parameters, i.e., extraction time, seaweed to water ratio, enzyme concentration, and extraction efficiency. Generally, the obtained results show that the protein extraction efficiency after enzymatic extraction was significantly higher when using a 1:20 ratio of seaweed to water, an extraction time of 6 h, and enzyme concentrations of 0.2 or 0.4% *w*/*w*. Shorter extraction time also resulted in some degree of protein extraction, but with lower extraction efficiency. The treatment with the number of 11, 19, 20 and 24 had the highest extraction efficiencies. However, treatment number 19, which had the ratio 1:20 and was conducted for 6 h with 0.2% *w*/*w* of each enzyme, was selected as the optimal condition. This result was in accordance with the previous results reported by Naseri et al. in which applied enzyme-assisted extraction was the best among the tested to extract protein from red seaweed *Palmaria palmata* [5].

3.2. Different Enzymes for Protein Extraction

In general, seaweeds proteins are bound by other non-protein components such as polyphenols and polysaccharides within the cell. Furthermore, seaweeds proteins may be found within macro-molecular assemblies or cross-linked through disulphide bonds to polysaccharides [5,6,22,27,28]. Some previous studies have demonstrated that applying alkaline solutions for example sodium hydroxide (NaOH) significantly improves the solubility and extraction of highly water-insoluble proteins from seaweeds and microalgae [5,27,29]. In addition, food-grade NaOH is used in the food industry, and as an example, in the extraction of protein-rich ingredients from different plants for example mainly soybean and chickpea. Two other enzymes (Viscozymes®, and Alcalase®) were also tested and compared with the enzymes already tested in step 1 (Celluclast®, Shearzyme®). Furthermore, all these enzymatic extractions were followed by the NAC-assisted alkaline extraction containing 1 g/L of NAC and 4 g/L of NaOH to test for possible further optimization. In order to evaluate the efficiency of NAC-assisted alkaline extraction alone, the first treatment was done with no added enzymes resulting in a protein extraction efficiency of 15.7%. When Celluclast® followed by NAC-assisted alkaline extraction was used, the results showed a steady increase from 19.4% at the concentration of 0.1% *w*/*w* enzyme to 37.9% for the concentration of 0.4% *w*/*w* enzyme. The results for Shearzyme® were different, with the lowest efficiency of 12.3% at the concentration of 0.2% *w*/*w*, while the highest efficiency was 35.8% at the lower enzymatic concentration of 0.1% *w*/*w*. Moreover, it was noticeable that due to an unexplainable reason, the obtained results for protein extraction efficiency of Shearzyme® at 0.2% *w*/*w* was lower than the treatment with no enzymes added.

The Viscozymes® treatment resulted in a significant increase in extraction efficiency in comparison to Celluclast® and Shearzyme®. The highest protein extraction efficiency was 48.5% at the concentration of 0.2% *w*/*w*. Viscozymes® is a multi-enzyme complex containing a wide range of carbohydrases, including arabinose, cellulase, β-glucanase, hemicellulase, and xylanase, and this most likely explains why it was more efficient than the other carbohydrate degrading enzymes.

In the current study, the protease Alcalase® was selected for testing in comparison with other non-protease enzymes, and the protein extraction efficiency of 59.4% was the significantly highest efficiency among all the experiments conducted in this study (concentration of 0.2% and tested at pH 7.0) (Table 2). Applying Alcalase® led to the hydrolysis of peptide bonds that link amino acids together in the polypeptide chain forming the protein, and this is most likely the reason for the significantly improved protein extraction. It has also been reported that the protein extraction efficiency could be increased if there was no limitation and pre-defined framework with respect to pH and temperature [4,5,23]. For example, the extraction efficiency for *Palmaria palmata* could reach up to 90% by using the combination of Alcalase® and Shearzyme® or Alcalase® and Celluclast® at pH 8 and the concentration of 0.2% for each one at a temperature of 50 °C for 14 h [5]. However, as mentioned before, all experiments of this study were done at room temperature (<30 °C) and at neutral pH in order to the adverse effect on carrageenan quality and to avoid gelling during the extraction process.

Table 2. Protein extraction efficiency (%) of the seaweed *Eucheuma denticulatum* treated with different enzymes followed by NAC-assisted alkaline extraction. Results are given as avearge ± standard deviation. Different superscripts letters in the same column indicate significant difference at the confidence level of 95% ($\alpha = 0.05$; $n = 3$).

Samples	Efficiency %
Only enzymes (Celluclast + Shearzyme (0.2%) pH = 7)—No NAC-assisted alkaline extraction	15.1 ± 0.53 [e,f]
No enzymes pH = 7	15.7 ± 0.12 [e,f]
Celluclast (0.1%) pH = 7	19.4 ± 0.49 [d,e,f]
Celluclast (0.2%) pH = 7	22.1 ± 8.89 [d,e]
Celluclast (0.4%) pH = 7	37.8 ± 2.73 [c]
Shearzyme (0.1%) pH = 7	35.8 ± 2.91 [c]
Shearzyme (0.2%) pH = 7	12.3 ± 1.34 [f]
Shearzyme (0.4%) pH = 7	24.8 ± 2.78 [d]
Viscozymes (0.1%) pH = 7	36.0 ± 8.14 [c]
Viscozymes (0.2%) pH = 7	48.5 ± 6.67 [b]
Viscozymes (0.4%) pH = 7	41.8 ± 2.60 [b,c]
Celluclast + Shearzyme (0.1%) pH = 7	20.5 ± 1.88 [d,e,f]
Celluclast + Shearzyme (0.2%) pH = 7	39.1 ± 7.39 [c]
Celluclast + Shearzyme (0.4%) pH = 7	24.8 ± 2.94 [d]
Celluclast + Shearzyme (0.2%) pH = 6	17.7 ± 3.48 [d,e,f]
Alcalase (0.2%) pH = 7	59.4 ± 1.41 [a]

Different letters indicate significant differences ($p < 0.05$).

3.3. *Effect of the Protein Extraction Process on the Yield of Carrageenan Extraction*

All the carrageenan extractions were performed based on an optimized method in the lab-scale done by Rhein-Knudsen et al. [25]. In the current study, the yield of isolated carrageenan extraction for the blank sample (with no enzymatic treatment and no NAC-assisted alkaline extraction) was 17.8%, while it increased to 23.8% when only NAC-assisted alkaline extraction with no enzymatic treatment was tested for protein extraction. Moreover, four different enzymatic treatments were selected to be compared with the blank samples, including Viscozymes (0.2% *w/w*) or Alcalase (0.2% *w/w*) or Celluclast (0.2% *w/w*) plus Shearzyme (0.2% *w/w*) at pH 6.0 and 7.0 followed by NAC-assisted alkaline extraction.

Table 3 shows that Viscozymes (0.2% *w/w*) or Alcalase (0.2% *w/w*) followed by NAC-assisted alkaline extraction had the lowest carrageenan isolation yield with 27.6 and 27.7%, respectively. Celluclast 0.2% *w/w* plus Shearzyme 0.2% *w/w* followed by NAC-assisted alkaline extraction at pH 6.0, and the same treatment at pH 7.0, had the significantly highest carrageenan yields. The results showed no significant difference between the two samples when treated by only Viscozymes (0.2% *w/w*) or Alcalase combined with NAC-assisted alkaline extraction ($p < 0.05$).

Table 3. Effect of protein extraction on the yield (%) of isolated carrageenan of seaweed *Eucheuma denticulatum* treated with different enzymes followed by NAC-assisted alkaline extraction. Yields are given in average ± standard deviation. Different superscripts letters in the same column indicate significant difference at the confidence level of 95% ($\alpha = 0.05$; $n = 3$).

Sample Description	Yield (%)
Blank (No enzymatic treatment, No NAC-assisted alkaline extraction)	17.8 ± 0.89^{d}
No enzymatic treatment, only NAC-assisted alkaline extraction	23.8 ± 1.62^{c}
Celluclast 0.2% + Shearzyme 0.2% (pH 6.0) + NAC-assisted alkaline extraction	$31.9 \pm 4.52^{a,b}$
Celluclast 0.2% + Shearzyme 0.2% (pH 7.0) + NAC-assisted alkaline extraction	35.5 ± 2.12^{a}
Viscozymes 0.2% + NAC-assisted alkaline extraction	$27.6 \pm 0.97^{b,c}$
Alcalase 0.2% + NAC-assisted alkaline extraction	$27.8 \pm 3.26^{b,c}$

Different letters indicate significant differences ($p < 0.05$).

Varadarajan et al. compared the use of a cellulase from *Aspergillus niger*, and traditional boiling on the extraction of carrageenan from *Eucheuma cottonii* [9] They achieved the highest carrageenan yield when using the cellulase (45% by weight) compared to 37% for fungal treated and 37.5% for the traditional extraction methods. However, the viscosity of the cellulase-extracted carrageenan was lower than the one extracted by the traditional method. The decrease in viscosity could be explained by the presence of impurities bound to the carrageenan as the cellulase attacks the cell walls in the seaweeds, to release the carrageenan, and thus does not degrade the carrageenan structure itself [21].

3.4. *Effect of the Protein Extraction Process on the Carrageenan Quality*

Table 4 shows the results for gel strength, turbidity, intrinsic viscosity (IV) and Food Chemicals Codex (FCC) viscosity of 1.5% solution for the isolated carrageenan from the blank and different samples obtained after protein extraction.

The obtained results for gel strength indicated that in comparison with the blank sample, enzymatic treatment with Celluclast® plus Shearzyme® did not change gel strength in 4, 8 and 12 mm whereas the enzymatic treatment with single use of Alcalase® or Viscozymes® increased the gel strength significantly (Table 4). These treatments had the highest protein extraction efficiency; hence, the lowest amount of protein was left in the sample, and therefore, the carrageenan is most likely of higher purity.

The maximum gel strength (breaking strength) and breaking distance for the samples were significantly lower for the combination of Celluclast® plus Shearzyme® at pH 6 and 7 compared to the untreated sample. The same was the case for the sample treated with NAC only. In contrast, there was no significant difference between the samples treated by only Alcalase® or Viscozymes® and the blank sample (Table 4).

Regarding the turbidity, there was a considerable deviation in the obtained results. The highest turbidity was for the blank sample, but it was evident that all the treated samples, especially those treated by the combination of Celluclast® plus Shearzyme®, had lower turbidity. The lowest turbidity was observed in the sample treated by Celluclast® 0.2% *w*/*w* plus Shearzyme® 0.2% *w*/*w* at pH 7 and followed by NAC-assisted alkaline extraction. Low turbidity is an advantage as it increases the possible applications of the carrageenan product. Therefore, protein extraction could be beneficial for expansion of carrageenan applications in different industries.

Intrinsic viscosity (IV) and FCC-viscosity as two parameters related to the gel viscosity indicated that there was no significant difference between the blank samples, and the samples treated by the combination of Celluclast® plus Shearzyme® or Viscozymes® for protein extraction. There was also no statistical difference between the samples treated by Alcalase®, or by only NAC.

Table 4. Effect of protein extraction of seaweed *Eurcheuma denticulatum* on the quality of carrageenan gel. Quality is given as averages ± standard deviations. Different superscript letters in the same column indicate significant difference at the confidence level of 95% ($\alpha = 0.05$; $n = 3$).

Sample Treatments	Gel Strength 4 mm (g/cm)	Gel Strength 8 mm (g/cm)	Gel Strength 12 mm (g/cm)	Breaking Strength (g/cm)	Breaking Distance (mm)	Turbidity (NTU *)	IV	FCC ** Viscosity
Blank (No enzymatic treatment, No NAC-assisted alkaline extraction)	$8.32 \pm 0.13^{b,c}$	15.6 ± 0.29^{b}	$23.6 \pm 0.50^{b,c}$	56.5 ± 3.53^{a}	26.5 ± 0.22^{a}	62.8 ± 25.2^{a}	8.03 ± 0.22^{a}	249 ± 24.7^{a}
NO enzymatic treatment, Only NAC-assisted alkaline extraction, pH = 7	8.09 ± 0.01^{c}	14.7 ± 0.14^{c}	21.5 ± 0.57^{c}	36.9 ± 6.71^{b}	22.0 ± 2.33^{c}	$43.8 \pm 11.2^{a,b,c}$	7.83 ± 0.51^{a}	233 ± 33.9^{a}
Celluclast 0.2% + Shearzyme 0.2% + NAC-assisted alkaline extraction, pH = 6	8.11 ± 0.42^{c}	14.9 ± 0.61^{c}	21.4 ± 0.01^{c}	34.6 ± 1.62^{b}	22.2 ± 0.92^{c}	$33.1 \pm 12.0^{b,c}$	7.51 ± 0.31^{a}	231 ± 49.5^{a}
Celluclast 0.2% + Shearzyme 0.2% + NAC-assisted alkaline extraction, pH = 7	8.53 ± 0.01^{b}	15.7 ± 0.14^{b}	22.6 ± 0.43^{c}	35.2 ± 1.50^{b}	21.9 ± 0.11^{c}	29.6 ± 5.72^{c}	7.59 ± 0.24^{a}	239 ± 24.0^{a}
Viscozyme 0.2% + NAC-assisted alkaline extraction, pH = 7	11.6 ± 0.20^{a}	19.1 ± 0.32^{a}	25.4 ± 2.95^{b}	59.0 ± 1.33^{a}	$24.6 \pm 0.60^{a,b}$	$48.1 \pm 2.14^{a,b,c}$	7.89 ± 0.02^{a}	222 ± 2.1^{a}
Alcalase 0.2% + NAC-assisted alkaline extraction, pH = 7	11.7 ± 0.31^{a}	19.4 ± 0.62^{a}	28.3 ± 1.52^{a}	59.1 ± 6.41^{a}	24.5 ± 0.80^{b}	$54.2 \pm 18.0^{a,b}$	7.48 ± 0.20^{a}	218 ± 14.8^{a}

NTU: Nephelometric Turbidity Unit, ** FCC: Food Chemicals Codex. Different letters indicate significant differences ($p < 0.05$).

In summary, the obtained results in the present study indicate that Alcalase® at 0.2% *w/w* and pH 7 or Viscozymes® at 0.2% *w/w* and pH 7 are the optimal treatment to extract protein from spinosum. As mentioned before, the highest extraction efficiencies were for Alcalase® at 0.2% *w/w* and pH 7, which led to a protein extraction efficiency of 59.4% and for Viscozymes® at 0.2% *w/w* and pH 7, the extraction efficiency was 48.5%. It was obvious that in comparison with the Viscozymes®, the use of Alcalase® will produce proteins/peptides with low molecular weight. Moreover, although the carrageenan yield for the combination of Celluclast® plus Shearzyme® was higher than when Viscozymes® or Alcalase® were used, the gel quality and in particular, the maximum gel strength (breaking strength) was lower for the combination of Celluclast® plus Shearzyme® compared to Viscozymes® or Alcalase®. The results in Table 4 showed that by using Celluclast® plus Shearzyme®, there was a negative effect on breaking strength compared to the blank sample. This was in compliance with the study performed by Varadarajan et al. (2009) [21].

Therefore, it could be concluded that using Alcalase® at 0.2% *w/w* and pH 7 or Viscozymes® at 0.2% *w/w* and pH 7 gave the higher protein extraction efficiency with no detrimental effects on carrageenan quality.

3.5. Comparison of Amino Acid Profiles

In the current study, the amino acid profile for the protein extracted by isoelectric precipitation at pH 3.5 for three different treatments were analyzed and compared, including Viscozymes® 0.2% *w/w* or Alcalase® 0.2% *w/w* or the combination of Celluclast® 0.2% *w/w* plus Shearzyme® 0.2% *w/w* which were followed by NAC-assisted alkaline extraction at pH 7. The reason for selecting these samples was in order to compare the sample with the highest carrageenan extraction yield with the samples with the highest protein extraction efficiencies.

The obtained results showed that protein extracted from the sample treated by the combination of Celluclast® 0.2% *w/w* and Shearzyme® 0.2% *w/w* had the highest amount of total amino acids with 50.8 mg/g DM. This was due to higher level of glutamic acid, aspartic acid, arginine and cysteine when compared to the protein extracted from the samples treated by Viscozymes® or Alcalase®. The protein extracted from samples treated by Viscozymes® 0.2% *w/w* or Alcalase® 0.2% *w/w* had 36.2 and 25.5 mg/g DM, respectively (Table 5). The amino acids of highest amount were glutamic acid, aspartic acid, cystine, leucine, alanine, and valine. In addition, in the sample treated with Viscozymes®, the lysine content was higher than for the other treatments. However, it was noticeable that arginine was not detectable in the samples treated with Viscozymes® or Alcalase®.

In comparison with the previous results obtained by the authors of this study with red seaweed *Palmaria palmata*, a protein extracted from *Euchema denticulatum* had a lower value of total amino acid ($\sum$AA) and total essential amino acids ($\sum$EAA) while the value for the ratio of EAA/AA was higher in protein extracted from *E. denticulatum*. The ratio was 27–42% for *P. palmata* while it was 46–54% for *E. denticulatum* [5]. The main reason could be due to the difference between the protein content of these two seaweeds.

The EAA/AA ratio of 46–54% of the protein of the *E. denticulatum* pellets in this experiment is in the same range as 48% EAA/AA in beef as a model organism for protein requirements, and the 47% EAA/AA in soybean [18,19]. The EEA/AA ratios calculated for beef and soybean were based on data for EEA and total protein content from Damodaran et al. [19]. In our calculation we have assumed that the protein content is equal to the sum of all amino acids. The branched chained amino acids (BCAA) are getting increasing attention, especially in fitness and body builder environments, since the leucine, isoleucine and valine are non-polar EAA that account for 35% of the EAA in the human muscles. The BCAA/AA ratio of the *E. denticulatum* pellets are 12–19% and therefore BCAA constitute a relative large part of the AA, which could make the product interesting for e.g., sport drinks and bars. However, it should be mentioned that protein content of beef and soybean is 18 and 40% respectively [19], compared to the 2.5–5.1% of the *E. denticulatum* pellets.

Table 5. Amino acid composition (mg amino acid/g DM), total essential amino acid content (∑ EAA), the essential amino acid ratio (EAA/AA) of the untreated samples of *Eucheuma denticulatum* seaweed, and extracted protein (pellet) obtained from the different selected enzymes used. The numbers are given as averages ± standard deviations of triplicates (n = 3).

Amino Acid	Untreated Seaweed	Viscozyme® L	Alcalase® 2.4 FL	Celluclast® 1.5 L + Shearzyme® 500 L
		Pellet	Pellet	Pellet
LYS [(1)]	1.12 ± 0.04	2.26 ± 0.01	1.18 ± 0.20	1.99 ± 0.12
ALA	1.01 ± 0.08	2.29 ± 0.62	1.47 ± 0.02	3.00 ± 0.48
ARG	0.64 ± 0.01	n.d.	n.d.	4.74 ± 0.23
C-C [(1)]	n.d	4.49 ± 1.03	6.00 ± 0.84	4.80 ± 0.06
MET [(1)]	0.41 ± 0.03	1.00 ± 0.16	0.61 ± 0.05	1.27 ± 0.01
LEU [(1)]	1.99 ± 0.27	2.64 ± 0.26	1.43 ± 0.01	3.35 ± 0.26
TYR [(1)]	0.43 ± 0.09	1.18 ± 0.12	0.73 ± 0.05	1.31 ± 0.07
PHE [(1)]	1.92 ± 0.07	2.08 ± 0.01	1.14 ± 0.10	2.59 ± 0.06
PRO	0.67 ± 0.05	1.48 ± 0.25	0.71 ± 0.04	1.93 ± 0.11
THR [(1)]	2.37 ± 0.04	1.14 ± 0.21	0.56 ± 0.04	1.56 ± 0.42
ASP	2.62 ± 0.23	3.38 ± 1.04	2.44 ± 0.13	5.06 ± 0.93
SER	1.48 ± 0.01	0.99 ± 0.28	0.57 ± 0.06	1.49 ± 0.31
HYP	n.d.	n.d.	n.d.	n.d.
GLU	2.64 ± 0.19	6.09 ± 2.25	4.68 ± 0.19	7.68 ± 1.73
VAL [(1)]	n.d	2.55 ± 0.26	1.10 ± 0.08	3.79 ± 0.89
HIS [(1)]	0.25 ± 0.03	0.55 ± 0.00	0.33 ± 0.07	0.62 ± 0.01
ILE [(1)]	1.08 ± 0.12	1.64 ± 0.03	0.57 ± 0.07	2.15 ± 0.18
GLY	0.89 ± 0.74	2.42 ± 0.59	1.97 ± 0.05	3.49 ± 0.42
∑ AA	19.5	36.2	25.5	50.8
∑ EAA	9.57	19.53	13.65	23.43
EAA/AA	0.49	0.54	0.54	0.46

[(1)] Essential amino acids for human nutrition. n.d. means not detected.

4. Conclusions and Future Perspectives

The present study successfully showed that it is possible to extract protein from *Eucheuma denticulatum,* in a multi-extraction setup, adding a new possible future protein resource with amino acid profiles comparable to meat (beef) and whey proteins, and further utilization going towards 'no waste' of our industrial resources. The best enzymes were Alcalase® or Viscozymes® at 0.2% *w/w*, and the maximum efficiency was increased by up to 60% of protein for Alcalase®. The present study furthermore demonstrated that the protein extraction process did not have a detrimental effect on the isolated carrageenan in the downstream carrageenan processing. For some parameters, the carrageenan behaved better than carrageenan extracted without the pre-extraction of protein. However, further investigation is needed to evaluate the bioactivity of both proteins and carrageenan before and after extraction.

Author Contributions: A.N., C.J., J.J.P.S., T.E.P., K.M.H., J.L., and S.L.H. contributed to conceiving the experiments, A.N., C.J., J.J.P.S. and S.L.H. planned the experiments. A.N. and J.J.P.S. carried out the experiments, analyzed the data, and A.N. wrote the main draft of the manuscript. All authors read and revised the paper and approved the final manuscript.

Funding: This research received funding from GUDP (Green Development and Demonstration Programme) with the journal number 34009-15-1029 as a part of the VALSEA project.

Acknowledgments: The authors wish to acknowledge Inge Holmberg (DTU Food), Helle Bech Olsen (CP Kelco), and Mohammad Amin Mohammadifar (DTU Food) for their technical assistance. Furthermore, we thank Novozymes A/S (Bagsværd, Denmark) for kindly providing all tested enzymes.

Conflicts of Interest: The authors declare no financial or other conflicts of interest.

References

1. Qin, Y. *Bioactive Seaweeds for Food Applications*; Elsevier Ltd.: Eastbourne, UK, 2018.
2. Torres, M.D.; Flórez-Fernández, N.; Domínguez, H. Integral utilization of red seaweed for bioactive production. *Mar. Drugs* **2019**, *17*, 314. [CrossRef] [PubMed]
3. Imeson, A.P. Carrageenan and furcellaran. In *Handbook of Hydrocolloids*; Phillips, G.O., Williams, P.A., Eds.; Woodhead Publishing Limited: Cambridge, UK, 2009; pp. 164–185. ISBN 9781845695873.
4. Bleakley, S.; Hayes, M. Algal Proteins: Extraction, Application, and Challenges Concerning Production. *Foods* **2017**, *6*, 33. [CrossRef] [PubMed]
5. Naseri, A.; Marinho, G.S.; Holdt, S.L.; Bartela, J.M.; Jacobsen, C. Enzyme-assisted extraction and characterization of protein from red seaweed Palmaria palmata. *Algal Res.* **2020**, *47*, 101849. [CrossRef]
6. Naseri, A.; Holdt, S.L.; Jacobsen, C. Biochemical and Nutritional Composition of Industrial Red Seaweed Used in Carrageenan Production. *J. Aquat. Food Prod. Technol.* **2019**, *28*, 967–973. [CrossRef]
7. Rhein-Knudsen, N.; Ale, M.T.; Meyer, A.S. Seaweed hydrocolloid production: An update on enzyme assisted extraction and modification technologies. *Mar. Drugs* **2015**, *13*, 3340–3359. [CrossRef] [PubMed]
8. Dawczynski, C.; Schubert, R.; Jahreis, G. Amino acids, fatty acids, and dietary fibre in edible seaweed products. *Food Chem.* **2007**, *103*, 891–899. [CrossRef]
9. Kadam, S.U.; Álvarez, C.; Tiwari, B.K.; O'Donnell, C.P. Extraction of biomolecules from seaweeds. In *Seaweed Sustainability: Food and Non-Food Applications*; Elsevier Inc.: Amsterdam, The Netherlands, 2015; ISBN 9780124199583.
10. Blanco-Pascual, N.; Alemán, A.; Gómez-Guillén, M.C.; Montero, M.P. Enzyme-assisted extraction of κ/ι-hybrid carrageenan from Mastocarpus stellatus for obtaining bioactive ingredients and their application for edible active film development. *Food Funct.* **2014**, *5*, 319–329. [CrossRef]
11. Barba, F.J.; Grimi, N.; Vorobiev, E. New Approaches for the Use of Non-conventional Cell Disruption Technologies to Extract Potential Food Additives and Nutraceuticals from Microalgae. *Food Eng. Rev.* **2014**, *7*, 45–62. [CrossRef]
12. Fleurence, J. Seaweed proteins. In *Proteins in Food Processing*; Elsevier Science: Amsterdam, The Netherlands, 2004; pp. 197–213, ISBN 0849325366.
13. Kadam, S.U.; Álvarez, C.; Tiwari, B.K.; O'Donnell, C.P. Extraction and characterization of protein from Irish brown seaweed Ascophyllum nodosum. *Food Res. Int.* **2017**, *99*, 1021–1027. [CrossRef]
14. Kulshreshtha, G.; Burlot, A.S.; Marty, C.; Critchley, A.; Hafting, J.; Bedoux, G.; Bourgougnon, N.; Prithiviraj, B. Enzyme-assisted extraction of bioactive material from Chondrus crispus and codium fragile and its effect on Herpes simplex virus (HSV-1). *Mar. Drugs* **2015**, *13*, 558–580. [CrossRef]
15. Chemat, F.; Zill-e-Huma; Khan, M.K. Applications of ultrasound in food technology: Processing, preservation and extraction. *Ultrason. Sonochem.* **2011**, *18*, 813–835. [CrossRef] [PubMed]
16. Topuz, O.K.; Gokoglu, N.; Yerlikaya, P.; Ucak, I.; Gumus, B. Optimization of Antioxidant Activity and Phenolic Compound Extraction Conditions from Red Seaweed (Laurencia obtuse). *J. Aquat. Food Prod. Technol.* **2016**, *25*, 414–422. [CrossRef]
17. Joannes, C.; Sipaut, C.S.; Dayou, J.; Md.Yasir, S.; Mansa, R.F. The Potential of Using Pulsed Electric Field (PEF) Technology as the Cell Disruption Method to Extract Lipid from Microalgae for Biodiesel Production. *Int. J. Renew. Energy Res.* **2015**, *5*, 598–621.
18. Kadam, S.U.; Tiwari, B.K.; O'Donnell, C.P. Application of novel extraction technologies for bioactives from marine algae. *J. Agric. Food Chem.* **2013**, *61*, 4667–4675. [CrossRef] [PubMed]
19. Azevedo, G.; Torres, M.D.; Sousa-Pinto, I.; Hilliou, L. Effect of pre-extraction alkali treatment on the chemical structure and gelling properties of extracted hybrid carrageenan from Chondrus crispus and Ahnfeltiopsis devoniensis. *Food Hydrocoll.* **2015**, *50*, 150–158. [CrossRef]
20. Anisuzzaman, S.M.; Bono, A.; Ding, O.W. Effect of process conditions on the gel viscosity and gel strength of semi-refined carrageenan (SRC) produced from seaweed (Kappaphycus alvarezii). *J. King Saud Univ. Eng. Sci.* **2014**, *26*, 3–9.
21. Varadarajan, S.A.; Ramli, N.; Ariff, A.; Said, M.; Yasir, S.M. Development of high yielding carragenan extraction method from Eucheuma Cotonii using cellulase and Aspergillus niger. In Proceedings of Prosiding Seminar Kimia Bersama UKM-ITB VIII9, Bangi, Malaysia, 9–11 June 2009; pp. 461–469.

22. Harnedy, P.A.; FitzGerald, R.J. Extraction of protein from the macroalga Palmaria palmata. *LWT Food Sci. Technol.* **2013**, *51*, 375–382. [CrossRef]
23. Bjarnadóttir, M.; Aðalbjörnsson, B.V.; Nilsson, A.; Slizyte, R.; Roleda, M.Y.; Hreggviðsson, G.Ó.; Friðjónsson, Ó.H.; Jónsdóttir, R. Palmaria palmata as an alternative protein source: Enzymatic protein extraction, amino acid composition, and nitrogen-to-protein conversion factor. *J. Appl. Phycol.* **2018**, *30*, 2061–2070. [CrossRef]
24. Angell, A.R.; Mata, L.; de Nys, R.; Paul, N.A. The protein content of seaweeds: A universal nitrogen-to-protein conversion factor of five. *J. Appl. Phycol.* **2016**, *28*, 511–524. [CrossRef]
25. Rhein-Knudsen, N.; Ale, M.T.; Ajalloueian, F.; Yu, L.; Meyer, A.S. Rheological properties of agar and carrageenan from Ghanaian red seaweeds. *Food Hydrocoll.* **2017**, *63*, 50–58. [CrossRef]
26. Sabeena Farvin, K.H.; Baron, C.P.; Nielsen, N.S.; Jacobsen, C. Antioxidant activity of yoghurt peptides: Part 1-in vitro assays and evaluation in ω-3 enriched milk. *Food Chem.* **2010**, *123*, 1081–1089. [CrossRef]
27. Amano, H.; Noda, H. Proteins of Protoplasts from Red Alga Porphyra yezoensis. *Nippon Suisan Gakkaishi* **1990**, *56*, 1859–1864. [CrossRef]
28. Miranda, C.F. Screening of Ionic Liquids for the Extraction of Proteins from the Macroalga Ulva Lactuca Aiming at an Integrated Biorefinery Approach. Master's Thesis, Instituto Superior Técnico, Lisboa, Portugal, 2017.
29. Lourenço, S.O.; Barbarino, E.; De-Paula, J.C.; Pereira, L.O.D.S.; Lanfer Marquez, U.M. Amino acid composition, protein content adn calculation of nitrogen-to-protein conversion factors for 19 tropical seaweeds. *Phycol. Res.* **2002**, *50*, 233–241. [CrossRef]

foods

MDPI

Article

Optimization of the Emulsifying Properties of Food Protein Hydrolysates for the Production of Fish Oil-in-Water Emulsions

Marta Padial-Domínguez *, F. Javier Espejo-Carpio, Raúl Pérez-Gálvez, Antonio Guadix and Emilia M. Guadix

Department of Chemical Engineering, University of Granada, 18010 Granada, Spain; fjespejo@ugr.es (F.J.E.-C.); rperezga@ugr.es (R.P.-G.); aguadix@ugr.es (A.G.); eguadix@ugr.es (E.M.G.)

* Correspondence: marthapadial@ugr.es; Tel.: +34-958241329

Received: 8 April 2020; Accepted: 14 May 2020; Published: 15 May 2020

Abstract: The incorporation of lipid ingredients into food matrices presents a main drawback—their susceptibility to oxidation—which is associated with the loss of nutritional properties and the generation of undesirable flavors and odors. Oil-in-water emulsions are able to stabilize and protect lipid compounds from oxidation. Driven by consumers' demand, the search for natural emulsifiers, such as proteins, is gaining much interest in food industries. This paper evaluates the in vitro emulsifying properties of protein hydrolysates from animal (whey protein concentrate) and vegetal origin (a soy protein isolate). By means of statistical modelling and bi-objective optimization, the experimental variables, namely, the protein source, enzyme (i.e., subtilisin, trypsin), degree of hydrolysis (2–14%) and emulsion pH (2–8), were optimized to obtain their maximal in vitro emulsifying properties. This procedure concluded that the emulsion prepared from the soy protein hydrolysate (degree of hydrolysis (DH) 6.5%, trypsin) at pH 8 presented an optimal combination of emulsifying properties (i.e., the emulsifying activity index and emulsifying stability index). For validation purposes, a fish oil-in-water emulsion was prepared under optimal conditions, evaluating its physical and oxidative stability for ten days of storage. This study confirmed that the use of soy protein hydrolysate as an emulsifier stabilized the droplet size distribution and retarded lipid oxidation within the storage period, compared to the use of a non-hydrolyzed soy protein isolate.

Keywords: emulsifying properties; statistical modelling; optimization; protein emulsifiers; physical stability; oxidative stability

1. Introduction

Oil-in-water emulsion systems are found in numerous natural and processed foods, such as milk, salad dressings, ice cream or butter, among others [1]. Besides, emulsions are increasingly used as delivery systems to encapsulate, protect and release functional ingredients into a food matrix. Such an example is omega-3 fortified fish oils [2,3]. The main drawback of incorporating omega-3 fatty acids as ingredients is their oxidative instability, so oil-in-water emulsions and encapsulation processes are a vehicle to protect them [4,5].

Additionally, these food emulsions are thermodynamically unstable complex systems and they can separate into their watery and oily phase over time [6]. Due to this fact, it is necessary to have the presence of certain components called emulsifiers with amphiphilic properties, which can interact in the oil–water interface and reduce the surface tension [7]. Emulsifiers are surfactants that facilitate the formation of the emulsion and preserve its stability over time [8]. Besides maintaining the physical stability of the emulsion over time, a good emulsifier acts as a barrier against lipid oxidation [9]. Thus, the amount of emulsifier adsorbed on the water–oil interface and its structure are factors to consider regarding the physical stability of the emulsions, and the oxidative stability of the lipids contained in them, during their storage [10].

The food industry employs proteins, phospholipids and other surfactants from natural or synthetic origin as emulsifiers [11]. Driven by consumer demand for natural products, there is an increasing interest in searching for natural emulsifiers from protein sources [12]. Food proteins from either animal (e.g., caseins, whey proteins) or plant origin (e.g., soy proteins) are amphiphilic molecules able to be absorbed at the interface of the emulsion and to stabilize the oil droplets [13]. The functional properties of such proteins are mainly determined by the position and number of amino acids in the polypeptide chain [14,15]. To this regard, a good emulsifier must present an optimal balance between polar and non-polar groups for an adequate water solubility and surface activity [16], which in turn has an important effect on the physical and oxidative stability of the emulsions. Accordingly, a change in the pH of the environment alters the net charge of the protein [17] and its conformation at the interface [18], affecting their emulsifying capacities.

Besides pH, the enzymatic hydrolysis of proteins has proved to improve their solubility and emulsification capacity due to the changes in the globular structure [19]. The extent of the proteolysis is accompanied by a decrease of the molecular weight and increase of the surface activity of the peptides [20], which favors the stability and viscoelasticity of the interfacial films in the emulsions [21]. However, an extensive hydrolysis is detrimental for the emulsifying properties, since short-chain peptides cannot reorient and unfold like large molecules to stabilize emulsions [22]. For instance, some authors reported that the emulsifying properties of whey protein hydrolysates improve in the range of degree of hydrolysis between 3 and 10% [23,24], while further proteolysis had an adverse effect. Considering the specificity of proteolytic enzymes, the type of enzyme employed also influences the in vitro emulsifying properties of the resulting hydrolysates. To this regard, Rdsa et al. [25] investigated the effect of different commercial proteases (i.e., subtilisin, trypsin, flavourzyme) in the hydrolysis of whey proteins, reporting significant differences in the emulsifying and antioxidant activities of the hydrolysates.

Therefore, the emulsifying capacity of peptides derived from the enzymatic hydrolysis of proteins depends on the protein source and the hydrolysis conditions, such as degree of hydrolysis (DH), type of enzyme and pH of the environment [26,27]. The degree of hydrolysis is defined as the percentage of total peptide bonds that are cleaved in the course of the enzymatic reaction. Although the effect of pH and DH on emulsion stability is reviewed extensively in the literature, a limited number of studies evaluate the impact of those processing variables on the oxidative stability of fish oil-in-water emulsions. The objective of this work is to model the emulsifying properties of whey and soy protein hydrolysates according to the enzyme, the final DH of the hydrolysates as well as the pH of the emulsion preparation, and test them in real emulsions of fish oil in water.

2. Materials and Methods

2.1. Protein Substrates and Enzymes

Whey protein concentrate (WPC) and soy protein isolate (Soy) were obtained from Wheyco GmbH, (Hamburg, Germany) and Solae LLC, (St. Louis, MO, USA), respectively. Subtilisin (EC 3.4.21.62) and trypsin (EC 3.4.21.4), kindly donated by Novozymes (Bagsvaerd, Denmark), were the enzymes employed in hydrolysis. The refined fish oil Omevital 18/12 TG Gold was purchased from BASF Personal Care and Nutrition GmbH (Illertissen, Germany) with a minimum content of omega-3 fatty acids of 35% (18% of EPA and 12% of DHA).

2.2. Hydrolysis Procedure

Hydrolysates of whey and soy protein were produced using subtilisin (EC 3.4.21.62) and trypsin (EC 3.4.21.4) (Novozymes, Denmark) as enzymes. To evaluate the effect of degree of hydrolysis on the emulsifying properties, five hydrolysates of a different degree of hydrolysis (2%, 4%, 6%, 10%

and 14%.) were produced with each enzyme. Hydrolysis was carried out in a jacketed stirred tank reactor that maintained a temperature at 50 °C during operation. When the protein solution (40 g/L) reached 50 °C, enzyme was added to the reactor at an enzyme/substrate ratio of 0.5%. The pH was maintained constant at pH 8 by adding NaOH 1M. The reaction was monitored by the pH-stat method [28], which relates in real time the consumption of base with the number of peptide bonds cleaved according to Equation (1):

$$DH = \frac{B \cdot N_b}{\alpha \cdot m_p \cdot h_{TOT}} \cdot 100 \tag{1}$$

where B (mL) is the amount of base consumed, N_b (eq/L) is the normality of the base, α is the average degree of dissociation of the α-NH_2 groups released during hydrolysis, m_p (g) is the mass of protein in the substrate and h_{TOT} (meq/g) is the number of equivalents of peptides bonds per gram of protein. After reaching the desired DH, the reaction was stopped by thermal deactivation of the enzyme at 90 °C for 5 min. Afterward, the samples were freeze-dried for further use.

2.3. Determination of the Emulsifying Properties

For each hydrolysate sample, the emulsifying activity and stability were determined at pH 2, 4, 6 and 8. Emulsifying activity is defined as the maximum amount of oil that can be emulsified by a fixed amount of the protein, while stability of the emulsion is defined as the rate of phase separation in water and oil during storage of the emulsion [29].

Hydrolysates samples were dissolved in ultrapure water and the pH was adjusted with 1 M NaOH or 1 M HCl. Then, 51 g of the aqueous solution at a protein concentration of 0.5% was mixed with 9 g of vegetable oil (i.e., sunflower oil) and then homogenized using an Ultra Turrax (IKA Werke GmbH &.Co., Staufen, Germany) at 16,000× rpm for 2 min. Two aliquots of the emulsion (50 μL) were pipetted at 0 and 10 min and diluted with 5 mL of a 0.1% sodium dodecyl sulphate (SDS) solution. The absorbance was determined with a 10-mm path length cuvette in a spectrophotometer (Cary 100 Bio, Varian, Palo Alto, CA, USA) at a wavelength 500 nm. The emulsifying activity index (EAI) and the emulsion stability index (ESI) were calculated by Equations (2) and (3), respectively:

$$EAI\left(\frac{m^2}{g}\right) = \frac{2 \cdot 2.303 \cdot A_0 \cdot DF}{(1-\theta)\ \cdot m_P} \tag{2}$$

$$ESI\ (min) = \frac{A_0 \cdot \Delta t}{A_0 - A_{10}} \tag{3}$$

where A_0 and A_{10} was the absorbance measured at initial time and after Δt = 10 min, respectively. The variables m_P and θ stand for the mass of protein (g) and the volume fraction of oil in the emulsion and DF is the dilution factor.

2.4. Statistical Modelling and Optimization

2.4.1. Statistical Design and Regression Model

The effects of substrate (i.e., WPC or soy protein), enzyme (i.e., subtilisin or trypsin), final degree of hydrolysis and pH of the emulsion were related to the emulsifying properties (i.e., emulsion activity and emulsion stability indices) of the hydrolysates by means of a statistical regression model. To this end, the experimental factors were varied according to a full factorial design, involving both categorical and numerical variables. Both substrate (noted as S) and enzyme (E) are categorical variables, each one taking two levels (i.e., WPC or soy protein for S; subtilisin or trypsin for E). Emulsion pH and the final degree of hydrolysis (DH) are numerical continuous variables, which were tested at 4 levels (pH 2, 4, 6 and 8) and 5 levels (DH 2%, 4%, 6%, 10% and 14%), respectively. Therefore, the full factorial design comprised 2 (S) × 2 (E) × 4 (pH) × 5 (DH) = 80 experimental trials. For each experimental run, the emulsion activity index (EAI, m^2/g) and the emulsion stability index (ESI, min) were determined as numerical outputs.

The nature of the input variables suggested employing a regression model with mixed categorical and continuous factors. Categorical variables can be incorporated into a regression model as numerical inputs by choosing an appropriate coding method. Among the different alternatives described in literature, the dummy coding method was chosen for this work [30,31]. By this approach, the levels of substrate were coded as 0 (soy protein isolate) and 1 (whey protein concentrate). As for the enzyme, it took the values 0 (trypsin) and 1 (subtilisin).

Based on the trend showed by EAI and ESI against pH and DH (Figures 1 and 2), a full cubic model was proposed for the multiple regression. The full cubic model, expressed by Equation (4), comprise linear, quadratic and cubic terms for the numerical variables, as well as their binary and ternary interactions with the categorical variables.

$$\begin{aligned}\{EAI, ESI\} = b_0 + b_1 \cdot S + b_2 \cdot E + b_3 \cdot pH + b_4 \cdot DH + b_{12} \cdot S \cdot E + b_{13} \cdot S \cdot pH + b_{14} \cdot S \cdot DH + \\ + b_{23} \cdot E \cdot pH + b_{24} \cdot E \cdot DH + b_{33} \cdot pH^2 + b_{44} \cdot DH^2 + b_{123} \cdot S \cdot E \cdot pH + b_{124} \cdot S \cdot E \cdot DH + \\ + b_{133} \cdot S \cdot pH^2 + b_{134} \cdot S \cdot pH \cdot DH + b_{144} \cdot S \cdot DH^2 + b_{233} \cdot E \cdot pH^2 + b_{234} \cdot E \cdot pH \cdot DH + \\ + b_{244} \cdot E \cdot DH^2 + b_{333} \cdot pH^3 + b_{334} \cdot pH^2 \cdot DH + b_{344} \cdot pH \cdot DH^2 + b_{444} \cdot DH^3\end{aligned} \quad (4)$$

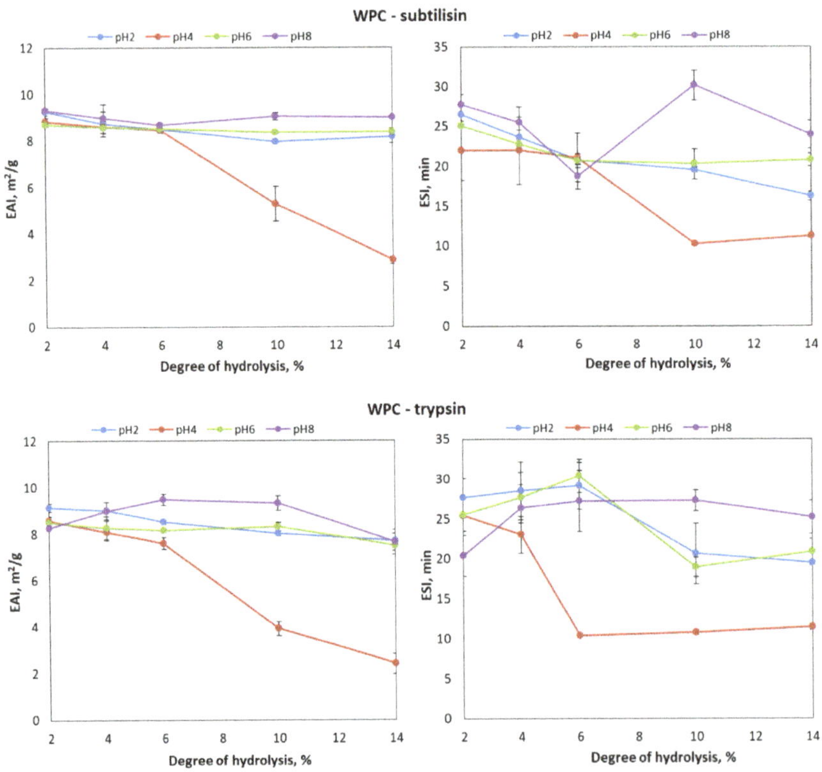

Figure 1. Emulsifying activity index (EAI, m^2/g) and emulsifying stability index (ESI, min) against pH (2, 4, 6 and 8) and degree of hydrolysis (2%, 4%, 6%, 10% and 14%) for the whey protein concentrate (WPC) hydrolysates produced with subtilisin and trypsin.

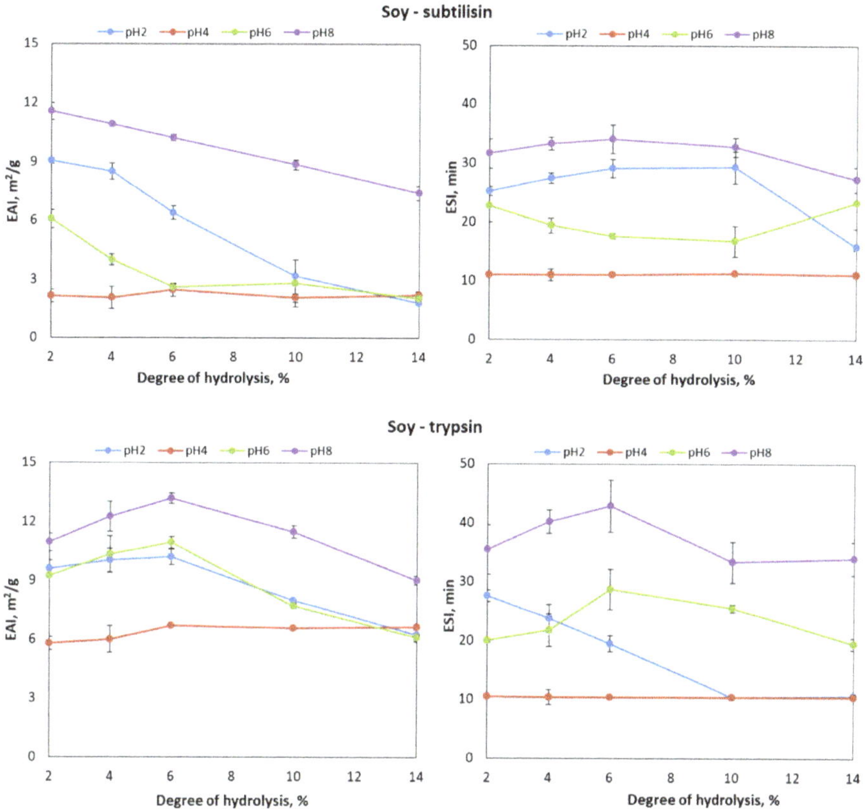

Figure 2. Emulsifying activity index (EAI, m^2/g) and emulsifying stability index (ESI, min) against pH (2, 4, 6 and 8) and degree of hydrolysis (2%, 4%, 6%, 10% and 14%) for the soy hydrolysates produced with subtilisin and trypsin.

The intercept and coefficients of the full cubic model were estimated by multiple regression. An ANOVA analysis was performed to evaluate the significance of each term of the regression model. An associated probability (p-value) was computed for each effect (i.e., linear, quadratic, cubic or interaction term in the polynomial) at a confidence level of 95%. Those effects with p-values above 0.05 were not considered as statistically significant on the responses of EAI or ESI, and could be removed from the full cubic model. The backward selection method [32] was chosen for this purpose. This approach consists of sequentially removing those effects with a higher p-value until obtaining a final reduced model, where all the terms are significant (i.e., p-value < 0.05) on the response variable.

2.4.2. Optimization of the Emulsifying Properties

The regression models expressed by Equation (4) are third-degree polynomials, which can be optimized to obtain a maximum for each response studied (EAI, ESI). The maximization of each response variable is desirable to ensure that the hydrolysate is able to stabilize the emulsion within the storage period. According to the experimental data, the operating conditions maximizing EAI may be detrimental for the optimization of ESI. The conflicting behavior of the experimental factors suggested combining both objectives into a single function OBJ (S, E, pH, DH) according to the weighted-sum method [33]. In our case, the optimization problem was stated as finding the combination of factors (i.e., S, E. pH, DH) within their experimental range, which maximizes the objective function:

$$\begin{array}{l} maximize\ OBJ(S,E,pH,DH) = w \cdot \frac{EAI}{\lambda_1} + (1-w) \cdot \frac{ESI}{\lambda_2} \\ subjected\ \text{to}: \\ \qquad S,E \in \{0,1\} \\ \qquad 2 \le pH \le 8 \\ \qquad 2 \le DH \le 14 \end{array} \quad (5)$$

The coefficient w is the weight factor, which represents the relative importance attached to a given single objective. For instance, $w = 1$ poses the single optimization of the emulsion activity index. In our case, where no single objective is predominant, $w = 0.5$. The responses EAI and ESI were normalized within the interval (0,1) by dividing by the coefficients λ_1 and λ_2, which are the maximal values predicted by the cubic models for the ESI (12.7 m^2/g) and EAI (39.0 min). The evolutionary algorithm, implemented in the Solver Tool of the MS Excel software, was chosen for the optimization.

2.5. Emulsion Preparation and Storage Study

The hydrolysate with the best emulsifying properties was selected for studying the stability of the emulsion. The native protein without hydrolysis was also used for producing a control emulsion. Emulsifiers (hydrolysate and native protein) were dissolved in distilled water and the pH was adjusted to the optimum identified using the models previously created. Emulsions containing 2% (*w*/*w*) protein and 5% (*w*/*w*) fish oil were produced similarly to García-Moreno et al. [5]. Initially, a pre-emulsion was prepared using Ultra Turrax (IKA Werke GmbH &.Co., Staufen, Germany) at 16,000× rpm for 2 min. The oil was added slowly to the aqueous phase during the first minute. Then, the homogenization was done in a high-pressure homogenizer (Panda Plus 2000, GEA Niro Soavi, Lübeck, Germany) at 450/75 bar, running 3 passes. The temperature during the process was controlled and kept under 26 °C. To accelerate lipid oxidation, a solution of 100 mM $FeSO_4$ was added to the emulsions (4 µL per 1 g of emulsion). A solution of 0.0125 g/mL of sodium azide was also added to prevent spoilage during storage. Emulsions were stored in the dark at 20 °C for 10 days in 30 mL glass bottles. Each bottle contained approximately 10 mL of emulsion. Samples were taken at days 0, 1, 3, 6 and 10 for droplet size and oxidative stability measurements.

2.6. Physical Stability of the Emulsions

The droplet size distribution was measured using a Malvern Mastersizer 2000 (Malvern Instruments Ltd., Worcestershire, UK). The emulsion samples were diluted in recirculating water (3000× rpm) until it reached an obscuration of 12–15%. The refractive indices of sunflower oil (1.469) and water (1.330) were used for particle and dispersant, respectively. Measurements were done in triplicate. The Sauter diameter D(3,2) was chosen as the mean diameter of the droplet size distribution.

The zeta potentials of the emulsions were estimated in a Zetasizer Nano ZS (Malvern Instruments Ltd., Worcestershire, UK). Emulsions were diluted 1:1000 in distilled water. Measurements were done in triplicate.

2.7. Oxidative Stability of Emulsions

To evaluate the oxidative stability of the emulsified oil, the hydroperoxide content and anisidine index were measured during the 10-day storage (0, 1, 3, 6 and 10 days). The extractions of the oil were done with a mixture of 2-propanol/hexane (1:1, *v*/*v*). The hydroperoxide value (PV) was determined using the thiocyanate assay [34]. Briefly, the extracted oil was diluted with 2-propanol and mixed with iron-II-chloride and ammonium thiocyanate solutions. Samples were incubated for 30 min at 60 °C; then, the absorbance was measured at 485 nm. Extractions were done in duplicate and the samples were done in quadruplicate. The presence of secondary oxidation compounds was determined by the anisidine value (AV), which was determined according to the standard ISO 6885:2006. The AV method is based on the reaction of p-anisidine diluted in acetic acid with the α and β unsaturated aldehydes present in the oil. Results were expressed as 100 times the increment of absorbance, measured at a wavelength of 350 nm in a 10 mm cell of the test solution when reacted with *p*-anisidine under the test conditions specified in the International Standard.

The total oxidation value (TOTOX) is a comprehensive oxidation index calculated from the weighted sum of the peroxide value (PV) and *p*-anisidine value (AV) as follows:

$$TOTOX = 2 \cdot PV + AV \quad (6)$$

2.8. Statistical Analysis

The analysis of variance (ANOVA) was carried out using Statgraphics (version 5.1. Statgraphics.Net., Madrid, Spain) Mean values were compared using the Tukey's multiple range test. Differences between means were considered significant at $p \leq 0.05$.

3. Results and Discussion

3.1. Characterization and Emulsifying Properties

Figure 1 is a plot of the in vitro emulsifying activity index (EAI) and the stability index (ESI) against DH (2–14%) and pH (2–8) for the fish oil-in-water emulsions that were stabilized with WPC and hydrolyzed with subtilisin and trypsin. For both enzymes, the levels of EAI slightly decreased with DH, presenting an average value between 8 and 10 m^2/g. This worsening with increasing DH was more pronounced for the emulsion at pH 4, where EAI decreased by 40%. As for the ESI values, they fluctuated along the interval of DH, presenting maximal values at DH 10% and pH 8 (subtilisin) and DH 6% and pH 6 (trypsin). These results are in line with Scherze and Muschiolik (2001) [24], who reported optimal emulsifying properties for oil-in-water emulsions stabilized with WPC hydrolysates at DH 10%. The stability of this emulsion was maintained even with hydrolysates at DH 20% and DH 27%.

Overall, the emulsions stabilized with soy hydrolysates presented maximal ESI values at DH 6%. Both the formation and stability indices worsened at DH levels above this value, regardless of the enzyme employed. These results agree with Jung, Murphy and Johnson (2005) [35], who reported maximal stability for the emulsions prepared with soy hydrolysates at DH 4% and pH 7. Similarly, Lopes-da-Silva and Monteiro (2019) [36] studied the effect of DH on the emulsifying properties of soy protein isolate, concluding that the emulsifying activity was superior at low DH levels.

The emulsifying properties of hydrolysates are mostly affected by the amphiphilicity of the peptides. It is widely reported that large peptides above 2 kDa improve the emulsifying properties of hydrolysates due to their ability to unfold at the oil/water interface [37]. Furthermore, they are more likely to have both hydrophobic and hydrophilic residues that interact with the oil droplets and the aqueous phase, respectively. This interaction increases the stability of the emulsion due to steric effects [13,38]. On the other hand, small-chain peptides delivered by enzymatic hydrolysis have a partially exposed hydrophobic core, which explains its higher diffusion rate at the oil/water interface and its ability to cover a larger area of the interface [19,38–40]. However, a large extent of the proteolysis yields a final hydrolysate with reduced amphiphilicity, which is detrimental for the emulsion formation and stability [13,41,42].

As for the influence of pH, Figures 1 and 2 show that the lowest EAI and ESI values were observed at pH 4, close to the isoelectric point of soy and whey proteins, where their solubility is lower [29,43–45]. Overall, pH values away from the isoelectric point improved the stability of the emulsion. This is confirmed by Figures 1 and 2, which show maximal EAI and ESI values for the emulsions at pH 8, regardless of the protein substrate and enzyme employed.

According to Figure 1, the behavior of the EAI against DH and pH was similar for WPC hydrolysates obtained with subtilisin and trypsin. In contrast, the emulsifying properties of soy protein hydrolysates produced with trypsin were significantly superior. Indeed, the maximum EAI (13.17 ± 0.27 m^2/g) and ESI (42.83 ± 4.39 min) values of the experimental series were found for soy protein hydrolyzed with trypsin at DH 6% (Figure 2). In line with these results, Zhao and Hou [46] reported that soy protein hydrolysates (DH 1–2%) produced with trypsin exhibited a better EAI than those hydrolyzed by neutrase. Similarly, other studies revealed the superior solubility and emulsifying properties of soy protein hydrolysates produced with trypsin [38,43]. This is explained by the specificity of trypsin towards bonds involving lysine or arginine carboxylic groups, releasing hydrophilic peptides that contribute to the stabilization of the oil droplets in the emulsion [46,47]. This behavior is more marked at low levels of DH, while at a DH above 8% their contribution to interphase hydrophobicity is similar to that provided by subtilisin [48].

3.2. Modelling and Optimization of the Emulsifying Properties

The emulsion activity and stability indices were modeled as a function of the type of the substrate and enzyme employed, as well as the degree of hydrolysis and pH. As mentioned above, the full cubic models (Equation (4)) were reduced by the backward selection method until obtaining the reduced models, where the terms with associated p-value lower than 5% were removed, as expressed by Equations (7) and (8):

$$\begin{aligned} EAI = {} & 21.093 - 4.827 \cdot S - 8.503 \cdot PH + 2.600 \cdot S \cdot pH - 1.794 \cdot E \cdot pH + 3.810 \cdot E \cdot S \\ & +1.572 \cdot pH^2 + 0.014 \cdot DH^2 - 0.080 \cdot pH^3 - 0.304 \cdot S \cdot pH^2 + 0.184 \cdot E \cdot pH^2 \end{aligned} \quad (7)$$

$$\begin{aligned} ESI = {} & 73.186 + 10.276 \cdot E - 38.974 \cdot pH + 7.346 \cdot S \cdot pH - 1.562 \cdot S \cdot DH - 2.077 \cdot E \cdot pH \\ & +0.135 \cdot pH \cdot DH - 13.219 \cdot E \cdot S + 7.390 \cdot pH^2 - 0.065 \cdot DH^2 - 0.390 \cdot pH^3 \\ & -1.029 \cdot S \cdot pH^2 + 0.084 \cdot S \cdot DH^2 + 2.403 \cdot E \cdot S \cdot pH \end{aligned} \quad (8)$$

The goodness-of-fit of the reduced model was confirmed by the coefficient of determination (R^2 = 81.4 and 83.0% for the EAI and ESI, respectively). The optimization problem stated by Equation (5) was applied for the WPC and soy protein isolate separately, in order to obtain the hydrolysis conditions (enzyme, DH) and emulsion pH for a maximal EAI or ESI. All these optimal solutions are summarized in Table 1.

Table 1. The optimal substrate, enzyme and hydrolysis parameters (pH, DH) for the maximal EAI, ESI or combined objective.

Objective	* S	* E	pH	* DH, %	*EAI, m^2/g	* ESI, min
Optimal EAI	WPC	Subtilisin	8	2	10.4	26.3
	Soy	Trypsin	8	2	12.7	36.5
Optimal ESI	WPC	Trypsin	2	2	8.8	29.7
	Soy	Trypsin	8	8.5	11.8	39.0
Optimal combined properties (EAI + ESI)	Soy	Trypsin	8	6.5	12.2	38.3

* S: Substrate; E: Enzyme; DH: Degree of Hydrolysis; EAI: Emulsifying Activity Index; ESI: Emulsifying Stability Index.

Since both single objectives (i.e., maximization of EAI and ESI) cannot be pursued simultaneously, they were combined into an objective function as expressed by Equation (5). The experimental factors maximizing the bi-objective function are also included in the Table 1.

Soy hydrolysates exhibited better emulsifying properties than the WPC hydrolysates. As for the enzymatic treatment, the emulsifying properties of the trypsin hydrolysates were generally superior to those obtained employing subtilisin as catalyst. As an exception, the maximal EAI of WPC (10.4 m^2/g) was predicted for the subtilisin hydrolysate at DH 2%. Both emulsifying indices were maximal at

DH 2%. The improvement of the emulsifying properties at a low DH is widely reported in the literature [13,37,38]. In contrast, the predicted emulsion stability index for soy hydrolysates (39 min) was maximal at DH 8.5%.

Overall, we can conclude that a pH of 8 led to improved activity and stability indices, except for the emulsion prepared from WPC hydrolyzed with trypsin. In this case, the emulsion presented optimal in vitro stability under acidic conditions (pH 2).

According to the optimization procedure, the emulsion at pH 8 prepared from soy hydrolysate with trypsin and DH 6.5% presented an optimal combination of process conditions to obtain the best EAI and ESI indices. The emulsifying indices predicted for this emulsion were 12.2 m^2/g and 38.3 min. These conditions were validated experimentally by a storage study.

The regression models expressed by Equations (6) and (7) allowed the generation of the surface plots for EAI and ESI. Figure 3 plots both properties for the soy trypsin hydrolysates as a function of pH and the degree of hydrolysis. The absolute maximums for the EAI and ESI are represented on the surface as a black marker (▾), while the white markers (▿) point out the intermediate solution optimizing both emulsifying properties.

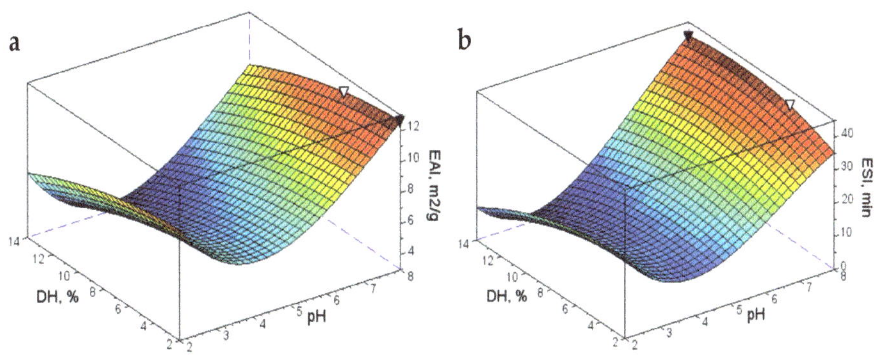

Figure 3. Surface plots of (**a**) EAI and (**b**) ESI for the soy hydrolysates.

As observed, both in vitro emulsifying properties were influenced similarly by the pH of the aqueous phase, attaining maximal levels at pH 8. Under neutral and alkaline conditions, increasing DH negatively affected the emulsion activity, while it significantly improved the emulsion stability. For both properties, a minimum was detected within the pH range 4–5, regardless the degree of hydrolysis, corresponding to the isoelectric pH for soy protein [45,46].

3.3. *Study of Physical and Oxidative Stability of the Emulsion*

3.3.1. Physical Stability of the Emulsion under Optimal Conditions

The optimization procedure described above predicts an optimal emulsion at pH 8, stabilized with hydrolysates of the soy protein isolate (DH 6.5%) hydrolyzed with trypsin. For validation purposes, an emulsion was prepared under optimal conditions (coded as SPH6.5) and subjected to a storage study, where the physical and oxidative stability were monitored over ten days at 20 °C. A control emulsion (SPI), prepared with non-hydrolyzed soy protein isolate, was subjected to the same storage study.

Both optimal and control emulsions presented high initial values of zeta potential, −52.57 ± 3.26 mV and −48.71 ± 2.40 mV, respectively. The negative potential is due to the pH of the emulsions, above the isoelectric point of the protein. The results are in line with other studies in emulsions with soy protein and its hydrolysates [49,50]. A higher absolute value of the zeta potential implies a greater stability of the emulsion [51] since it reflects a better repulsion between the oil droplets, and their aggregation is also minimized [52]. In general, with zeta potential values higher than 30 mV or less than −30 mV, aggregation particles is prevented due to their electrostatic repulsions [53].

Figure 4 shows the evolution of the droplet size distribution, represented by $D_{3,2}$ throughout the storage period. It is observed that the droplet size distribution of both emulsions was stable until the sixth day of storage. Indeed, the reference diameter $D_{3,2}$ increased continuously over time at a rate of 0.02 μm per day until Day 6. From this day on, droplet size underwent a slight increase of 0.07 μm up to 0.221 ± 0.01 μm (Figure 4). The control emulsion (SPI) presented a similar behavior during the storage period, showing a slight tendency towards a bimodal distribution on Day 10, where a second peak centered at D(3,2) = 7.4 μm was detected (data not shown). The coalescence of oil droplets into larger units shows physical destabilization of the emulsion system, which could be accompanied by increased lipid oxidation. As shown in Figure 4, the Sauter diameters for the optimal emulsion were significantly lower than those of the control emulsion throughout the storage period. We concluded that the hydrolysis of the soy protein isolate improved its emulsifying properties with respect to the native protein.

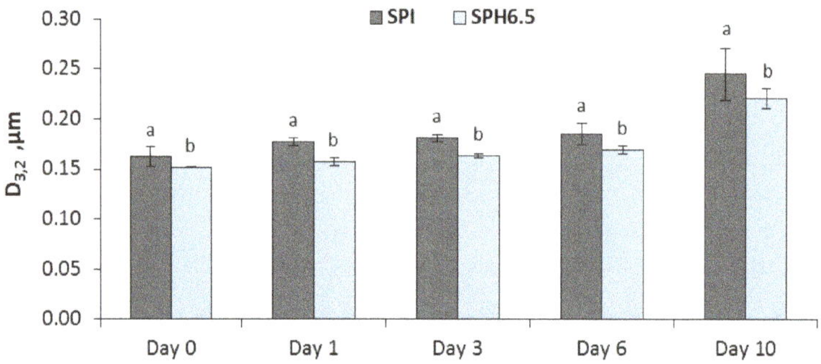

Figure 4. Droplet size of emulsions stabilized with a soy protein isolate (SPI) and soy protein hydrolysate (SPHT6.5) during storage at 20 °C. * Different letters represent significant differences between SPI and SPH6.5 ($p < 0.05$).

Similar results were obtained for an oil-in-water emulsion stabilized with rice protein hydrolyzed with trypsin (DH 6.36%) [54]. In this study, there was an increase of 0.04 μm in the droplet size during storage. On Day 1 of storage the $D_{3,2}$ was 0.27 ± 0.01 μm, and 0.31 ± 0.02 μm on Day 7 [54]. To achieve good physical stability, it is important that the fortified emulsions (i.e., emulsions enriched with Omega-3 fatty acids) have an average droplet size below 0.5 μm [3,55]. Previous studies with non-hydrolyzed vegetable protein report average droplet diameters from 0.3 μm [55–58] up to more than 1 μm [49,50].

3.3.2. Oxidative Stability of the Emulsion under Optimal Conditions

The peroxide value of the oil extracted from the emulsions SPI and SPH6.5 was monitored throughout the storage study (Figure 5a) in order to evaluate its oxidative stability. For the optimal SPH6.5 emulsion, the initial peroxide value (PV) is similar to that obtained for refined fish oil (<1 mmol O_2/kg oil). It is widely accepted that the emulsification process increases the susceptibility of oil droplets to lipid oxidation [6] since it increases the contact between the polyunsaturated fatty acids with oxygen and other pro-oxidative agents. In this regard, the interface acts as an effective barrier to the oxidative stability of the emulsion [59]. In our case, this effect could be enhanced by the emulsifying and antioxidant properties of the soy protein hydrolysate, which has been reported in the literature [60–66].

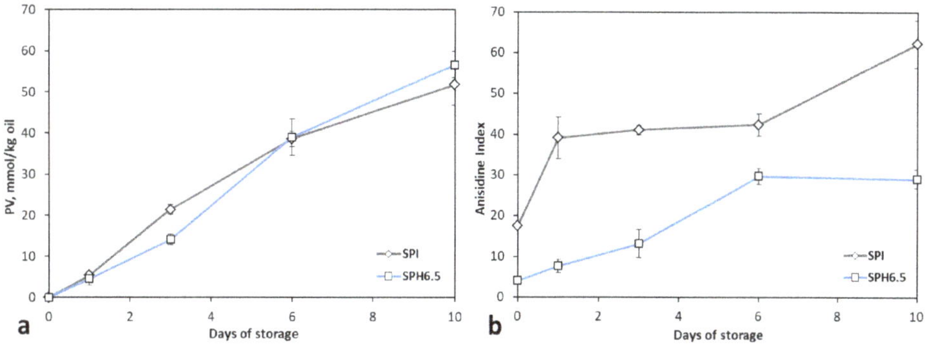

Figure 5. Evolution of (**a**) peroxide value (mmol/kg oil) and (**b**) anisidine index for the optimal (SPH6.5) and control (SPI) emulsions.

As shown in Figure 5a, we observed that the PV for the SPH6.5 emulsion showed a constant increase from the first day of storage, until reaching a value of 56.65 mmol/kg oil on Day 10. A similar trend was also found for the evolution of the concentration of primary oxidation compounds in rapeseed oil emulsion stabilized with fava bean hydrolysate [67], in a fish oil-in-water emulsion stabilized with soy β-conglycinin hydrolysate [68] and with lecithin, widely used as a natural emulsifier. In stabilized oil-in-water emulsions with lecithin, the peroxide concentration increased during longer storage [69]. The authors indicate that it is possible that endogenous Fe^{2+} or Fe^{3+} ions bound to the lecithin layer in the emulsion through electrostatic attraction between the negatively charged lecithin-coated droplets and the positively charged iron ions could promote the lipid oxidation of the emulsions [69]. Differences in the PV values between our work and the latter studies are due to the incubation process carried out in our analyses to determine PV.

The PV levels of the control emulsion (SPI) increased at a higher rate than SPH6.5 up to the third day of storage. From this point on, the values of the PV were slightly lower than those observed for SPH6.5. Finally, we observed no significant differences between emulsions at the end of the storage study. Although this trend could reflect higher stability of the SPI against lipid oxidation, it does not agree with the evolution shown by anisidine index (Figure 5b). Indeed, the SPI presented higher levels of anisidine values throughout the storage period and underwent a sharp increase above the sixth day. The conflicting behavior between the peroxide and anisidine values shown by the SPI sample can be explained by the progress of oxidation reactions. Lipid oxidation results in primary peroxide compounds, which are further decomposed into secondary products. The latter cannot be quantified by the PV analysis, but are reflected in the anisidine index. García-Moreno et al. [5] also reported this trend, concluding that the rate of decomposition of peroxides into secondary products was higher than the formation rate.

The higher oxidative stability shown by the soy protein hydrolysate (SPH6.5) is confirmed by the evolution of the Totox value (Figure 6), which quantifies the global generation of oxidation products during processing. According to this figure, the SPH6.5 sample exhibited higher stability against lipid oxidation throughout the storage period. The sharp increase between Days 3 and 6 is also reflected in this index, although the overall Totox values of the hydrolysate were inferior to those observed for the protein isolate (SPI).

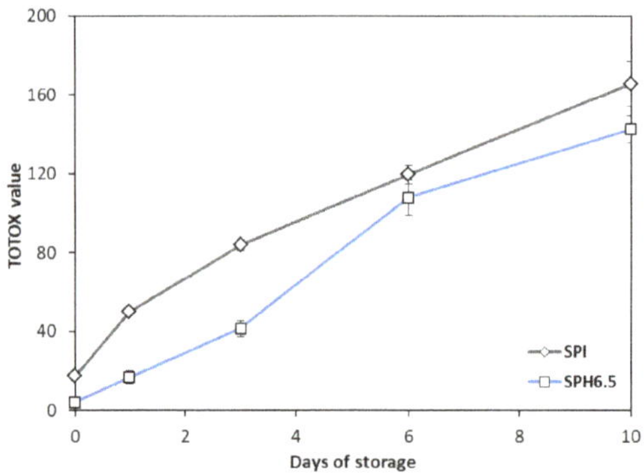

Figure 6. Evolution of Totox value for the optimal (SPH6.5) and control (SPI) emulsions.

The anisidine value (AV) measures the secondary oxidation compounds and it is more sensitive to aldehydes, principally 2-alkenals and 2, 4-alkadienals, present in the emulsified oil [70,71]. Emulsion control showed an AV significantly higher than optimal emulsion (Figure 5b). That could be due to the fact that hydrolysis partially releases hydrophobic groups of the protein and the hydrolysate could show efficient free radical scavenging capacity and iron chelating activity compared to the intact protein [63,72,73]. Peptides derived from hydrolysis, due to their structure, are less prone to conformational folding than proteins and this ensures that most of their reactive groups are always available for antioxidant reactions [40].

4. Conclusions

Protein hydrolysate with improved emulsifying properties were produced from different protein sources (whey and soy protein) using subtilisin and trypsin as enzymes. The effect of substrate, enzyme, degree of hydrolysis and the pH on the emulsifying properties of hydrolysates was assayed. Generally, soy hydrolysates seemed to present higher emulsifying properties than whey protein hydrolysates. However, because of the complexity of the data, a mathematical model including all the variables was proposed for optimizing the emulsifying activity index (EAI) and the emulsion stability index (ESI). A maximum EAI of 12.7 m^2/g was obtained for the soy hydrolysate at DH 2% produced by trypsin, while the maximum ESI (39 min) was identified for the soy hydrolysate at DH 8.5%. In order to undertake the bi-objective optimization problem, which involves both having conflicting results, the weighted-summed method generated an optimal solution, taking into account that both objectives have the same importance. According to this, the emulsion at pH 8 prepared from the soy hydrolysate with trypsin and a DH of 6.5% presented an optimal combination of emulsifying properties. The emulsion prepared under the optimal conditions proposed presented steady values of the Sauter diameter within the first six days of storage, always below those corresponding to the emulsion employing the soy protein isolate. Moreover, the hydrolysate was able to retard lipid oxidation from Day 0 to 6 of storage when compared to the use of the non-hydrolyzed soy protein isolate. In this line, these emulsions could be incorporated into certain food matrices, such as mayonnaise or salad dressings, to enrich them with omega-3 fatty acids.

Author Contributions: Conceptualization, F.J.E.-C., M.P.-D. and E.M.G.; methodology, M.P.-D. and F.J.E.-C.; software, F.J.E.-C. and R.P.-G.; writing—original draft preparation, M.P.-D. and R.P.-G.; supervision, F.J.E.-C., A.G. and E.M.G.; funding acquisition, A.G. and E.M.G. All authors have read and agreed to the published version of the manuscript.

Funding: This research was funded by the project CTQ2017-87076-R from the Spanish Ministry of Science and Innovation.

Conflicts of Interest: The authors declare no conflict of interest.

References

1. McClements, D.J. Lipid-Based Emulsions and Emulsifier. In *Food Lipids*; Akoh, C.C., Min, D.B., Eds.; CRC Press Taylor & Francis Group: Boca Raton, FL, USA, 2008; pp. 64–96.
2. McClements, D.J.; Li, Y. Structured emulsion-based delivery systems: Controlling the digestion and release of lipophilic food components. *Adv. Colloid Interface Sci.* **2010**, *159*, 213–228. [CrossRef] [PubMed]
3. Walker, R.; Decker, E.A.; McClements, D.J. Development of food-grade nanoemulsions and emulsions for delivery of omega-3 fatty acids: Opportunities and obstacles in the food industry. *Food Funct.* **2015**, *6*, 41–54. [CrossRef] [PubMed]
4. Barrow, C.J.; Wang, B.; Adhikari, B.; Liu, H. *Spray Drying and Encapsulation of Omega-3 Oils*; Woodhead Publishing Limited: Cambridge, UK, 2013; ISBN 9780857094285.
5. García-Moreno, P.J.; Guadix, A.; Guadix, E.M.; Jacobsen, C. Physical and oxidative stability of fish oil-in-water emulsions stabilized with fish protein hydrolysates. *Food Chem.* **2016**, *203*, 124–135. [CrossRef] [PubMed]
6. Mcclements, D.J. *Food Emulsion Principles, Practices, and Technique*, 2nd ed.; CRC Press: Boca Raton, FL, USA, 2005.
7. Jacobsen, C. Some strategies for the stabilization of long chain n-3 PUFA-enriched foods: A review. *Eur. J. Lipid Sci. Technol.* **2015**, *117*, 1853–1866. [CrossRef]
8. Ozturk, B.; McClements, D.J. Progress in natural emulsifiers for utilization in food emulsions. *Curr. Opin. Food Sci.* **2016**, *7*, 1–6. [CrossRef]
9. McClements, D.J. *Food Emulsions*, 3th ed.; CRC Press: Boca Raton, FL, USA, 2015; ISBN 9781498726696.
10. Klompong, V.; Benjakul, S.; Kantachote, D.; Shahidi, F. Antioxidative activity and functional properties of protein hydrolysate of yellow stripe trevally (Selaroides leptolepis) as influenced by the degree of hydrolysis and enzyme type. *Food Chem.* **2007**, *102*, 1317–1327. [CrossRef]
11. Uluata, S.; McClements, D.J.; Decker, E.A. Physical Stability, Autoxidation, and Photosensitized Oxidation of ω-3 Oils in Nanoemulsions Prepared with Natural and Synthetic Surfactants. *J. Agric. Food Chem.* **2015**, *63*, 9333–9340. [CrossRef]
12. Can Karaca, A.; Low, N.H.; Nickerson, M.T. Potential use of plant proteins in the microencapsulation of lipophilic materials in foods. *Trends Food Sci. Technol.* **2015**, *42*, 5–12. [CrossRef]
13. Lam, R.S.H.; Nickerson, M.T. Food proteins: A review on their emulsifying properties using a structure-function approach. *Food Chem.* **2013**, *141*, 975–984. [CrossRef]
14. Damodaran, S. Protein Stabilization of Emulsions and Foams. *J. Food Sci.* **2006**, *70*, R54–R66. [CrossRef]
15. Damodaran, S.; Parkin, K. *Fennema's Food Chemistry*; CRC Press: Boca Raton, FL, USA, 2017.
16. MClements, D.J.; Jafari, S.M. Improving emulsion formation, stability and performance using mixed emulsifiers: A review. *Adv. Colloid Interface Sci.* **2018**, *251*, 55–79. [CrossRef] [PubMed]
17. Kralova, I.; Sjöblom, J. Surfactants Used in Food Industry: A Review. *J. Dispers. Sci. Technol.* **2009**, *2691*, 1363–1383. [CrossRef]
18. McClements, D.J.; Gumus, C.E. Natural emulsifiers—Biosurfactants, phospholipids, biopolymers, and colloidal particles: Molecular and physicochemical basis of functional performance. *Adv. Colloid Interface Sci.* **2016**, *234*, 3–26. [CrossRef] [PubMed]
19. Jeewanthi, R.K.C.; Lee, N.-K.; Paik, H.-D. Improved Functional Characteristics of Whey Protein Hydrolysates in Food Industry. *Korean J. Food Sci. Anim. Resour.* **2015**, *35*, 350–359. [CrossRef]
20. Davis, J.P.; Doucet, D.; Foegeding, E.A. Foaming and interfacial properties of hydrolyzed β-lactoglobulin. *J. Colloid Interface Sci.* **2005**, *288*, 412–422. [CrossRef]
21. Tamm, F.; Sauer, G.; Scampicchio, M.; Drusch, S. Pendant drop tensiometry for the evaluation of the foaming properties of milk-derived proteins. *Food Hydrocoll.* **2012**, *27*, 371–377. [CrossRef]
22. Rahali, V.; Chobert, J.M.; Haertlé, T.; Guéguen, J. Emulsification of chemical and enzymatic hydrolysates of β-lactoglobulin: Characterization of the peptides adsorbed at the interface. *Nahrung-Food* **2000**, *44*, 89–95. [CrossRef]
23. Lieske, B.; Konrad, G. Physico-chemical and functional properties of whey protein as affected by limited papain proteolysis and selective ultrafiltration. *Int. Dairy J.* **1996**, *6*, 13–31. [CrossRef]
24. Scherze, I.; Muschiolik, G. Effects of various whey protein hydrolysates on the emulsifying and surface properties of hydrolysed lecithin. *Colloids Surfaces B Biointerfaces* **2001**, *21*, 107–117. [CrossRef]

25. Da Rdsa, L.D.L.; Santana, M.C.; Avezedd, T.L.; Brígoda, A.O.S.; Gdddy, R.; Pachecd, S.; Mellonger-Solva, C.; Cabral, L.M.C. A comparison of dual-functional whey hydrolysates by the use of commercial proteases. *Food Sci. Technol.* **2018**, *38*, 31–36.
26. Ipsen, R.; Otte, J.; Sharma, R.; Nielsen, A.; Gram Hansen, L.; Bruun Qvist, K. Effect of limited hydrolysis on the interfacial rheology and foaming properties of β-lactoglobulin A. *Colloids Surfaces B Biointerfaces* **2001**, *21*, 173–178. [CrossRef]
27. Chalamaiah, M.; Dinesh Kumar, B.; Hemalatha, R.; Jyothirmayi, T. Fish protein hydrolysates: Proximate composition, amino acid composition, antioxidant activities and applications: A review. *Food Chem.* **2012**, *135*, 3020–3038. [CrossRef] [PubMed]
28. Camacho, F.; González-Tello, P.; Páez-Dueñas, M.P.; Guadix, E.M.; Guadix, A. Correlation of base consumption with the degree of hydrolysis in enzymic protein hydrolysis. *J. Dairy Res.* **2001**, *68*, 251–265. [CrossRef] [PubMed]
29. Pearce, K.N.; Kinsella, J.E. Emulsifying Properties of Proteins: Evaluation of a Turbidimetric Technique. *J. Agric. Food Chem.* **1978**, *26*, 716–723. [CrossRef]
30. Davis, M.J. Contrast coding in multiple regression analysis: Strengths, weaknesses, and utility of popular coding structures. *J. Data Sci.* **2010**, *8*, 61–73.
31. Nagadevara, V. Advanced Regression Analysis. In *International Series in Operations Research and Management Science*; Springer New York LLC.: New York, NY, USA, 2019; Volume 264, pp. 247–281.
32. Kroese, D.P.; Chan, J.C.C. *Statistical Modeling and Computation*; Springer: New York, NY, USA, 2014.
33. Marler, R.T.; Arora, J.S. The weighted sum method for multi-objective optimization: New insights. *Struct. Multidiscip. Optim.* **2009**, *41*, 853–862. [CrossRef]
34. Drusch, S.; Serfert, Y.; Berger, A.; Shaikh, M.Q.; Rätzke, K.; Zaporojtchenko, V.; Schwarz, K. New insights into the microencapsulation properties of sodium caseinate and hydrolyzed casein. *Food Hydrocoll.* **2012**, *27*, 332–338. [CrossRef]
35. Jung, S.; Murphy, P.A.; Johnson, L.A. Physicochemical and functional properties of soy protein substrates modified by low levels of protease hydrolysis. *J. Food Sci.* **2005**, *70*, C180–C187. [CrossRef]
36. Lopes-da-Silva, J.A.; Monteiro, S.R. Gelling and emulsifying properties of soy protein hydrolysates in the presence of a neutral polysaccharide. *Food Chem.* **2019**, *294*, 216–223. [CrossRef]
37. Schröder, A.; Berton-Carabin, C.; Venema, P.; Cornacchia, L. Interfacial properties of whey protein and whey protein hydrolysates and their influence on O/W emulsion stability. *Food Hydrocoll.* **2017**, *73*, 129–140. [CrossRef]
38. Adjonu, R.; Doran, G.; Torley, P.; Agboola, S. Whey protein peptides as components of nanoemulsions: A review of emulsifying and biological functionalities. *J. Food Eng.* **2014**, *122*, 15–27. [CrossRef]
39. Sinha, R.; Radha, C.; Prakash, J.; Kaul, P. Whey protein hydrolysate: Functional properties, nutritional quality and utilization in beverage formulation. *Food Chem.* **2007**, *101*, 1484–1491. [CrossRef]
40. Aluko, R.E. Amino acids, peptides, and proteins as antioxidants for food preservation. In *Handbook of Antioxidants for Food Preservation*; Shahidi, F., Ed.; Elsevier Ltd.: Amsterdam, The Netherlands, 2015; pp. 105–140. ISBN 9781782420972.
41. Gbogouri, G.A.; Linder, M.; Fanni, J.; Parmenter, M. C: Food Chemistry and Toxicology Influence of Hydrolysis Degree. *J. Food Sci.* **2004**, *69g*, 615–622. [CrossRef]
42. Chen, L.; Chen, J.; Ren, J.; Zhao, M. Modifications of soy protein isolates using combined extrusion pre-treatment and controlled enzymatic hydrolysis for improved emulsifying properties. *Food Hydrocoll.* **2011**, *25*, 887–897. [CrossRef]
43. Taherian, A.R.; Britten, M.; Sabik, H.; Fustier, P. Ability of whey protein isolate and/or fish gelatin to inhibit physical separation and lipid oxidation in fish oil-in-water beverage emulsion. *Food Hydrocoll.* **2011**, *25*, 868–878. [CrossRef]
44. Joye, I.J.; McClements, D.J. Emulsifying and emulsion-stabilizing properties of gluten hydrolysates. *J. Agric. Food Chem.* **2014**, *62*, 2623–2630. [CrossRef]
45. Taheri, A.; Farvin, K.H.S.; Jacobsen, C.; Baron, C.P. Antioxidant activities and functional properties of protein and peptide fractions isolated from salted herring brine. *Food Chem.* **2014**, *142*, 318–326. [CrossRef]
46. Zhao, X.; Hou, Y. Limited hydrolysis of soybean protein concentrate and isolate with two proteases and the impact on emulsifying activity index of hydrolysates. *African J. Biotechnol.* **2009**, *8*, 3314–3319.

47. Ashaolu, T.J. Applications of soy protein hydrolysates in the emerging functional foods: A review. *Int. J. Food Sci. Technol.* **2019**, 421–428. [CrossRef]
48. Mutilangi, W.A.M.; Panyam, D.; Kilara, A. Functional Properties of Hydrolysates from Proteolysis of Heat-denatured Whey Protein Isolate. *J. Food Sci.* **1996**, *61*, 270–275. [CrossRef]
49. Wang, S.; Yang, J.; Shao, G.; Qu, D.; Zhao, H.; Yang, L.; Zhu, L.; He, Y.; Liu, H.; Zhu, D. Soy protein isolated-soy hull polysaccharides stabilized O/W emulsion: Effect of polysaccharides concentration on the storage stability and interfacial rheological properties. *Food Hydrocoll.* **2020**, *101*, 105490. [CrossRef]
50. Zhu, L.; Xu, Q.; Liu, X.; Xu, Y.; Yang, L.; Wang, S.; Li, J.; Ma, T.; Liu, H. Oil-water interfacial behavior of soy β-conglycinin–soyasaponin mixtures and their effect on emulsion stability. *Food Hydrocoll.* **2020**, *101*, 105531. [CrossRef]
51. Hou, P.; Pu, F.; Zou, H.; Diao, M.; Zhao, C.; Xi, C.; Zhang, T. Whey protein stabilized nanoemulsion: A potential delivery system for ginsenoside Rg3 whey protein stabilized nanoemulsion: Potential Rg3 delivery system. *Food Biosci.* **2019**, *31*, 100427. [CrossRef]
52. Wu, Z.; Wu, J.; Zhang, R.; Yuan, S.; Lu, Q.; Yu, Y. Colloid properties of hydrophobic modified alginate: Surface tension, ζ-potential, viscosity and emulsification. *Carbohydr. Polym.* **2018**, *181*, 56–62. [CrossRef] [PubMed]
53. Astete, C.E.; Sabliov, C.M.; Watanabe, F.; Alexandru, B. Ca2+ Cross-linked alginic acid nanoparticles for solubilization of lipophilic natural colorants. *J. Agric. Food Chem.* **2009**, *57*, 7505–7512. [CrossRef] [PubMed]
54. Pan, X.; Fang, Y.; Wang, L.; Xie, M.; Hu, B.; Zhu, Y.; Zhao, E.; Pei, F.; Shen, F.; Li, P.; et al. Effect of enzyme types on the stability of oil-in-water emulsions formed with rice protein hydrolysates. *J. Sci. Food Agric.* **2019**, *99*, 6731–6740. [CrossRef]
55. Gumus, C.E.; Decker, E.A.; McClements, D.J. Formation and Stability of ω-3 Oil Emulsion-Based Delivery Systems Using Plant Proteins as Emulsifiers: Lentil, Pea, and Faba Bean Proteins. *Food Biophys.* **2017**, *12*, 186–197. [CrossRef]
56. Hu, M.; McClements, D.J.; Decker, E.A. Lipid oxidation in corn oil-in-water emulsions stabilized by casein, whey protein isolate, and soy protein isolate. *J. Agric. Food Chem.* **2003**, *51*, 1696–1700. [CrossRef]
57. Qiu, C.; Zhao, M.; Decker, E.A.; McClements, D.J. Influence of protein type on oxidation and digestibility of fish oil-in-water emulsions: Gliadin, caseinate, and whey protein. *Food Chem.* **2015**, *175*, 249–257. [CrossRef]
58. Chung, C.; Sher, A.; Rousset, P.; Decker, E.A.; McClements, D.J. Formulation of food emulsions using natural emulsifiers: Utilization of quillaja saponin and soy lecithin to fabricate liquid coffee whiteners. *J. Food Eng.* **2017**, *209*, 1–11. [CrossRef]
59. C. Genot, T.-H.; Kabri, A.M. Stabilization of omega-3 oils and enriched foods using emulsifiers. In *Food Enrichment with Omega-3 Fatty Acids*; Jacobsen, C., Nielsen, N.S., Horn, A.F., Sørensen, A.-D.M., Eds.; Woodhead Publishing Limited: Cambridge, UK, 2013; pp. 151–193.
60. McClements, D.J.; Decker, E.A. Lipid oxidation in oil-in-water emulsions: Impact of molecular environment on chemical reactions in heterogeneous food systems. *J. Food Sci.* **2000**, *65*, 1270–1282. [CrossRef]
61. Zhao, J.; Xiong, Y.L. Interfacial peptide partitioning and undiminished antioxidative and emulsifying activity of oxidatively stressed soy protein hydrolysate in an O/W emulsion. *LWT-Food Sci. Technol.* **2015**, *61*, 322–329. [CrossRef]
62. Di Giorgio, L.; Salgado, P.R.; Mauri, A.N. Encapsulation of fish oil in soybean protein particles by emulsification and spray drying. *Food Hydrocoll.* **2019**, *87*, 891–901. [CrossRef]
63. Zhang, Y.; Niu, F.; Zhang, X.; Lu, Z.; Guo, Y.; Wang, H. Controlled enzymatic hydrolysis on characteristic and antioxidant properties of soybean protein isolate-maltodextrin conjugates. *Int. J. Food Prop.* **2018**, *21*, 2239–2249. [CrossRef]
64. Peña-Ramos, E.A.; Xiong, Y.L. Antioxidant Activity of Soy Protein Hydrolysates in a Liposomal System. *J. Food Sci.* **2002**, *67*, 2952–2956. [CrossRef]
65. Gibbs, B.F.; Zougman, A.; Masse, R.; Mulligan, C. Production and characterization of bioactive peptides from soy hydrolysate and soy-fermented food. *Food Res. Int.* **2004**, *37*, 123–131. [CrossRef]
66. Moure, A.; Domínguez, H.; Parajó, J.C. Antioxidant properties of ultrafiltration-recovered soy protein fractions from industrial effluents and their hydrolysates. *Process Biochem.* **2006**, *41*, 447–456. [CrossRef]
67. Liu, C.; Bhattarai, M.; Mikkonen, K.S.; Heinonen, M. Effects of Enzymatic Hydrolysis of Fava Bean Protein Isolate by Alcalase on the Physical and Oxidative Stability of Oil-in-Water Emulsions. *J. Agric. Food Chem.* **2019**, *67*, 6625–6632. [CrossRef]

68. Phoon, P.Y.; San Martin-Gonzalez, M.F.; Narsimhan, G. Effect of hydrolysis of soy β-conglycinin on the oxidative stability of O/Wemulsions. *Food Hydrocoll.* **2014**, *35*, 429–443. [CrossRef]
69. Ogawa, S.; Decker, E.A.; McClements, D.J. Influence of Environmental Conditions on the Stability of Oil in Water Emulsions Containing Droplets Stabilized by Lecithin–Chitosan Membranes. *J. Agric. Food Chem.* **2003**, *51*, 5522–5527. [CrossRef]
70. Fereidoon, S.; Udaya, N. Wanasundara Methods for Measuring Oxidative Rancidity in Fats and Oils. In *Food Lipids: Chemistry, Nutrition and Biotechnology*; Akoh, C.C., Min, D.B., Eds.; CRC PressTaylor & Francis Group: Boca Raton, FL, USA, 2008; pp. 387–408.
71. Jacobsen, C. Challenges when developing omega-3 enriched foods. *OCL-Ol. Corps Gras Lipides* **2010**, *17*, 251–258. [CrossRef]
72. Zhang, Y.; Tan, C.; Abbas, S.; Eric, K.; Zhang, X.; Xia, S.; Jia, C. The effect of soy protein structural modification on emulsion properties and oxidative stability of fish oil microcapsules. *Colloids Surfaces B Biointerfaces* **2014**, *120*, 63–70. [CrossRef] [PubMed]
73. Zhang, Y.; Tan, C.; Abbas, S.; Eric, K.; Xia, S.; Zhang, X. Modified SPI improves the emulsion properties and oxidative stability of fish oil microcapsules. *Food Hydrocoll.* **2015**, *51*, 108–117. [CrossRef]

foods

MDPI

Article

Reducing the High Iodine Content of *Saccharina latissima* and Improving the Profile of Other Valuable Compounds by Water Blanching

Cecilie Wirenfeldt Nielsen [1,2,*], Susan Løvstad Holdt [1], Jens J. Sloth [1], Gonçalo Silva Marinho [1], Maren Sæther [3], Jon Funderud [3] and Turid Rustad [2]

1 National Food Institute, Technical University of Denmark, 2800 Kgs. Lyngby, Denmark; suho@food.dtu.dk (S.L.H.); jjsl@food.dtu.dk (J.J.S.); goncalomarinho@hotmail.com (G.S.M.)

2 Department of Biotechnology and Food Science, Norwegian University of Science and Technology, 7018 Trondheim, Norway; turid.rustad@ntnu.no

3 Seaweed Energy Solutions AS, 7491 Trondheim, Norway; sather@seaweedenergysolutions.com (M.S.); funderud@seaweedenergysolutions.com (J.F.)

* Correspondence: cewini@food.dtu.dk; Tel.: +45-3196-4945

Received: 29 March 2020; Accepted: 30 April 2020; Published: 4 May 2020

Abstract: *Saccharina latissima* contains high amounts of iodine in comparison to other seaweeds. The present study aimed to decrease the iodine content of *S. latissima* (sugar kelp) by water blanching and freezing to avoid an excess intake of iodine by consumption of sugar kelp. Various blanching conditions were investigated (temperature; 30, 45, 60 and 80 °C, and duration; 2, 30, 120 and 300 s). Some conditions resulted in a significant decrease in iodine content (≥45 °C and ≥30 s). Non-processed *S. latissima* contained on average 4605 mg iodine kg^{-1} dw^{-1} which significantly decreased following the treatments. The lowest content obtained was 293 mg iodine $kg^{-1}{\cdot}dw^{-1}$ by water blanching at 80 °C for 120 s. The study also investigated if other valuable compounds were affected during the processing conditions. No significant changes were observed for total lipid and protein, but significant changes were seen for ash. A significant loss of two non-essential amino acids (glutamic acid and alanine) due to the blanching process was found. This also resulted in a protein quality increase as the essential amino acid to total amino acid ratio changed from 42.01 ± 0.59% in fresh seaweed to 48.0 ± 1.2% in blanched seaweed. Moreover, the proportion of eicosapentaenoic acid, α-linolenic acid, polyunsaturated fatty acids, and omega-3 fatty acids (%FAME), and the omega-3 to omega-6 fatty acids ratio was significantly higher in the samples blanched at 60 °C for 300 s compared to the fresh and samples blanched at 45 °C for 30 s. The total phenolic content (TPC) and the radical scavenging activity were significantly higher in treated samples. The results indicate that the processing did not compromise the valuable compounds in focus in this study for *S. latissima*; they did, however, result in biomass with an improved profile of health beneficial compounds.

Keywords: sugar kelp (*Saccharina latissima*); seaweed; blanching; freezing; iodine; nutrients; bioactives; antioxidant activity

---

1. Introduction

Seaweed as a food source is currently in focus in Europe due to its potential as an environmentally friendly and nutritious food source. It grows in seawater, does not take up any land areas and does not need any freshwater supply. Moreover, seaweeds contain highly valuable bioactive compounds [1], which are of interest from a nutritional point of view. *Saccharina latissima* (Linnaeus) is a brown macroalga commonly known as sugar kelp, which is successfully cultivated in Europe and is commercially available. Valuable compounds specifically for *S. latissima* are minerals, essential amino acids, polyunsaturated fatty acids, phenolic compounds, antioxidants, etc. [2–5]. Kelps, in general, contain high amounts of the trace element iodine [6] and contents as high as 6500 mg·kg^{-1}·dw^{-1} have been found in European sugar kelp [7]. Marine foods are considered rich in iodine and contain up to 30 mg·kg^{-1}·dw^{-1} [8]. It is clear that the iodine content of sugar kelp is extremely high.

Due to the possibilities of both insufficient and excess intake of iodine, dietary values for recommended intake (RI) (150 µg iodine day^{-1} for adults) and upper intake level (UL) (600 µg iodine day^{-1} for adults) have been established to provide guidance to consumers by both the Nordic Council of Ministers [9] and the European Food Safety Authorities (EFSA) [10,11]. No maximum levels for iodine in food (including seaweed) have been established in European Food Regulation. However, some member states, e.g., France, have published a recommended maximum level of 2000 mg iodine kg^{-1} dry seaweed product [12].

It is in the manufacturer's interest, as well as their responsibility, to ensure that their food products are safe and comply with food legislation (Council Directive (EC) 178/2002; [13]). The high iodine content of sugar kelp can, even from low consumption, lead to an intake of iodine above the upper level and hence, this is seen as a possible market barrier for the trade of sugar kelp. Consequently, the European seaweed industry demands procedures that can reduce the iodine content of their final products.

Previous studies have investigated how to reduce iodine in sugar kelp, e.g., by water soaking (low temperatures) or boiling [7,14]. Recently, Stévant et al. (2018) [7] found that soaking in water at 32 °C for 1–6 h would reduce the iodine content significantly by 84%–88%. Another study by Lüning and Mortensen (2015) [14] found a significant iodine reduction of 33% and 75% for sugar kelp boiled for 2 and 20 min, respectively.

The aim of this study was to investigate the possible iodine reduction by processing. Moreover, to investigate if the various processing conditions compromised other valuable compounds, the nutritional value of the final product was determined. The processing conditions investigated were water blanching at short processing times (2, 30, 120 and 300 s) at various water temperatures (30, 45, 60 and 80 °C) or by freezing followed by thawing. Moreover were the protein by sum of amino acids, amino acid profile, total lipid, fatty acid profile, ash, total phenolic content, and antioxidant capacity quantified to evaluate the possible quality compromise of *S. latissima* due to processing. In addition, the true retention factors were calculated in order to show not only the proximate composition, but also link to losses of biomass during processing. Lastly, a brief assessment of the iodine content compared to the recommended intake (RI) and upper intake levels (UL) was conducted.

2. Materials and Methods

2.1. Chemicals

All chemicals were of analytical grade unless otherwise stated. More specifically, tetra-methyl-ammonium-hydroxide (TMAH) 25%, sodium hydroxide, o-phtaldialdehyde and butylated hydroxytoluene were purchased from Sigma-Aldrich (Steinheim, Germany). Toluene, hydrogen chloride, methanol, chloroform, sodium chloride, and 20% boron trifluoride were purchased from Merck (Darmstadt, Germany). n-Heptane was purchased from VWR (Radnor, Pennsylvania, USA). The purified C23:0 was purchased from Nu Chek Prep (Elysian, MN, USA).

2.2. Raw Material

Saccharina latissima was harvested at Seaweed Energy Solution's cultivation site at Frøya, Norway (N63° 42.279′ E8° 52.232′). Approximately 1 m long blades were harvested on 23 April 2018. After harvest, the sugar kelp was stored in aerated tanks with flow-through seawater at 7–8 °C for 2–3 days until processing was carried out. Stem and holdfast were kept, but fouling organisms were removed by hand.

2.3. *Water Blanching and Freezing*

All treatments were performed on samples (n = 3) of 150 ± 20.0 g wet weight (*ww*) whole thallus sugar kelp. Water blanching was conducted in a JBN12 (Grant Instruments Ltd., England) water bath in 5 L tap water with the following variables: temperature 30, 45, 60 and 80 °C and processing durations of 2, 30, 120 and 300 s. The water bath was cleaned and the water renewed prior to each treatment process including between treatment replicates. After blanching, the sugar kelp was drained by keeping it vertical for 5–10 s, then placed in a zip lock plastic bag and cooled in ice water for 3 min.

Freezing was conducted in a −20 °C freezing room (Schneider Electric, Rueil-Malmaison, France) for 8 h and thawing was done at 5 °C overnight (freeze-thawed). The freeze-thaw caused drip water to appear, which was drained by keeping it vertical for 5–10 s.

From the harvest batch were three replicates of approximately 150 g fresh sugar kelp collected, drained and stored until analyses. These samples are referred to as "fresh sugar kelp".

2.4. *Sample Preparation for Chemical Analysis*

Prior to analysis, the sugar kelp samples were cut into 3 × 3 cm pieces and gently mixed. Approximately 20% of the material was used for water and ash analysis. The rest was freeze dried (Alpha 1-4 LDplus, Martin Christ, Germany) at −40 °C, and then homogenized by milling (MM 400, Retsch, Germany) to particle sizes of <300 µm.

2.5. *Dry Matter and Ash*

Dry matter content (DM) was determined gravimetrically by vaporizing water at 105 °C for 20–24 h in an oven (Termaks AS, Bergen, Norway) until stable weight [15]. Ash content was determined gravimetrically by ignition in a muffle furnace (Nabertherm, Lilienthal, Germany) at 600 °C for 15–20 h [16]. Both analyses were performed in duplicates.

2.6. *Iodine*

Inductively Coupled Plasma Mass Spectrometry (ICP-MS) was used for the quantification of the total iodine content in the sugar kelp samples. The samples were prepared according to EN17050:2017 [17]. Briefly, 0.15–0.20 g of freeze-dried milled homogenized powder was weighed into tubes (Kimax®). Subsequently, 5 mL Milli-Q® water and 1 mL 25% tetra-methyl-ammonium-hydroxide (TMAH) were added. The tubes were then sealed and placed in a preheated oven at 90 ± 3.0 °C for 3 h followed by cooling and diluting to a final volume of 20 mL with Milli-Q® water. To remove coarse particles, the samples were centrifuged at 10,000× g for 20 min. Prior to analysis, the supernatant was filtered through a 0.45 µm filter and samples were diluted 50 times. Sample extracts were stored in metal free plastic tubes for a maximum of 5 days prior to ICP-MS analysis. The iodine quantification (n = 1) was performed by ICP-MS (Finnigan ELEMENT-2, Thermo Fisher, Waltham, MA, USA) combined with an SC2 DX auto sampler and a prepFAST auto dilution system (Elemental Scientific, Omaha, NE, USA). The parameter settings were 15.5 L·min^{-1} coolant gas, 1.1 L·min^{-1} auxiliary gas, and 0.75 L·min^{-1} nebulizer gas. Isotopes monitored were 127I and 185Re for internal standard. The limit of quantification (LOQ) for iodine was 37 µg·g^{-1}. The certified reference material (NIST 3232, Kelp powder) was analyzed together with the samples and the obtained results complied well with the certified value (recovery 96.8%, n = 2).

2.7. *Amino Acid Hydrolysis and Calculation of Protein Content*

Briefly, 50 mg sample was hydrolyzed in 1 mL 6 M HCl at 105 °C for 22 h (n = 2). Prior to quantification, the samples were neutralized by NaOH and HCl to pH 7.0 ± 0.5 and filtered through a Whatman glass microfiber filter (GF/C) using suction. The samples were diluted 1:100 with distilled water followed by a 0.22 µm filtration. Then the amino acids were quantified by High-Performance Liquid Chromatography (HPLC) (Dionex UltiMate® 3000 HPLC+ focused, Dionex UltiMate® 3000 Autosampler, Dionex RF

Fluorescence Detector, Thermo Scientific, USA) including precolumn derivatization of the amino acids with o-phtaldialdehyde and Nova-Pak® column (C18, 4 μm). Tryptophan was destroyed in acid hydrolysis, thus not detected. The chromatographic peaks for glycine and arginine gathered in one, therefore an average of their molar masses was used to calculate their content.

The total protein content was calculated by summing the total moles of amino acids as recommended by Angell et al. (2016) [18] and FAO (2003) [19] and then subtracting the water mass (18 g H_2O mol^{-1} amino acid), which was integrated during disruption of peptide bonds in the acid hydrolysis [20].

2.8. Determination of Total Lipid Content

The gravimetric method described by Bligh and Dyer (1959) [21] was used to quantify total lipid content. Briefly, a mixture of demineralized water, methanol and chloroform (0.8:2:1 mL) was added to 30 mg freeze-dried sample followed by homogenizing with 1 mL chloroform (20 s) and then 1 mL demineralized water (20 s). The mixture was centrifuged thoroughly at 4 °C. The chloroform phase (0.5 mL) was added to a pre-weighed glass container and vaporized overnight in a fume hood. The following day the container was weighed again. The total lipid content was calculated by the following Equation:

$$\text{tatal lipid}(\%) = \frac{1v \cdot C \cdot 100}{Cv \cdot m} \tag{1}$$

where l_v is the lipid weighed after vaporization in mg, c is the added chloroform (2 mL), c_v is the vaporized chloroform (0.5 mL), and m is the mass of the weighed sample before extraction.

2.9. Carbohydrates by Difference

Estimation of the total carbohydrate content was done by "total carbohydrate by difference" [19], which includes fibers:

$$\text{carbohydrates} = 100 - (\text{weightinggram}\ [\text{protein} + \text{lipid} + \text{water ash}] \text{in } 100 \text{ g of food}) \tag{2}$$

2.10. Fatty Acids

Direct methylation of fatty acids were performed according to [22]. Approximately, 100 mg of sample was mixed with 1 mL 1.0 M NaOH in methanol, 1 mL toluene, and 0.1 mL 2% (*w*/*v*) C23:0 in *n*-heptane and sonicated for 10 min, followed by 100 °C water bath for 2 min and cooling in cold water. Next, 2 mL boron trifluoride (20% solution) were added and boiled and cooled as earlier described. Lastly, 2 mL 6.8 M saturated sodium chloride solution along with 1 mL heptane with 0.01% butylated hydroxytoluene were added and shaken. The heptane phase was transferred to a GC vial and FAMEs were analyzed by GC (HP 5890A, Agilent Technologies, Palo Alto, CA, USA) according to the American Oil Chemists' Society (AOCS) [23]. For separation, DB127-7012 column (10 m × ID 0.1 mm × 0.1 μm film thickness, Agilent Technologies, Palo Alto, CA, USA) was used. Injection volume was 0.2 μL in split mode (1:50). The initial temperature of the GC-oven was 160 °C. The temperature was set to increase gradually as follows: 160–200 °C (10.6 °C min^{-1}), 200 °C kept for 0.3 min, 200–220 °C (10.6 °C min^{-1}), 220 °C kept for 1 min, 220–240 °C (10.6 °C min^{-1}) and kept at 240 °C for 3.8 min. The determination was conducted in duplicates. Fatty acids were identified by comparison of retention times with those from a mixture of known fatty acid standards. Results were given in area %.

2.11. Extraction for Antioxidant Analyses and Total Phenolic Content

The extraction of antioxidants was executed according to [24] with some modifications. Briefly, 0.2 g sugar kelp sample was weighed, and 5 mL methanol was added, and the samples were placed in a sonicator for 30 min. Thereafter, the samples were centrifuged (2164× g for 10 min) and the supernatant was collected. The pellets were resuspended and extractions repeated twice. The solvent was evaporated under nitrogen flow. When the extracts were completely dry, they were stored in the freezer (−18 °C). Prior to the analyses, the dried powders were dissolved in 1 mL methanol.

2.12. Total Phenolic Content (TPC)

The procedure was carried out according to [25]. Extracts (100 µL) were mixed with Folin–Ciocalteu reagent (0.75 mL, 10% *v*/*v*). After 5 min, 0.75 mL sodium carbonate (7.5% *w*/*v*) was added to the mixture, which was then incubated at room temperature in darkness for 90 min. The absorbance was measured at 725 nm. The measured absorbance was converted into gallic acid equivalents by a standard curve of gallic acid in the range of 7.8–250 µg mL^{-1}.

2.13. DPPH Radical Scavenging Activity

The radical scavenging activity was performed according to the method described by Yang, Guo and Yuan (2008) [26] with modifications. Briefly, 100 µL of methanolic extracts were added to a microplate followed by 100 µL 0.1 mM DPPH soluted in methanol and mixed followed by incubation for 30 min in darkness at room temperature. The absorbance was measured at 517 nm in a microplate reader (Synergy 2 BioTek, Winooski, VT, USA). Triplicate measurements were performed and butylated hydroxytoluene (BHT) was included in the assay as a positive control since a concentration of 0.91 mM of BHT is giving approximately 70% inhibition. A sample blank was made with DPPH but without extract solution (Ab) and a sample control was made without DPPH but with extract/fraction solution (A0). Results are expressed as IC50, i.e., the concentration of extract needed to obtain 50% inhibition. The % DPPH radical scavenging activity was calculated as follows:

$$\text{DPPH radical scavenging activity} = \left(1 - \frac{\text{As} - \text{A0}}{\text{Ab}}\right) \times 100 \quad (3)$$

2.14. Mass Balances and True Retentions

All samples were weighed before and after treatment with one decimal accuracy. Before weighing, the samples were drained by keeping them vertical for 5–10 s. The true retention (TR) of a compound is the proportion of a particular nutrient that remains after processing relative to the original content of that specific nutrient. True retentions were calculated based on the proximate composition before and after processing and were calculated as suggested by [27]:

$$\text{TR} = \frac{\text{g retasined nutrient} \cdot \text{g total product post treatment}}{\text{g original nutrient} \cdot \text{g total product prior treatment}} \quad (4)$$

In cases where a replicate of a specific nutrient concentration was missing due to analytical mistakes, the missing replicate was interpolated from the other analytical replicates by taking an average of the known replicates for that specific treatment.

2.15. Statistical Analysis

The results are given as mean ± standard deviation. The statistical analyses were carried out in the software SPSS Statistics 24 (IBM Corp., Armonk, NY, USA). The test run to define the statistically significant difference between the means of the groups (fresh, freeze-thawed, and blanched material) was a one-way ANOVA with Tukey's post-hoc test. A one-way PERMANOVA was used to test the effect of processing on total phenolic content, and radical scavenging activity (PERMANOVA package in PRIMER+; [28]; type III sum of squares and unrestricted permutation (9999) on raw data; $\alpha = 0.05$) with a posteriori analysis (pairwise test). Means were considered statistically significantly different when levels of $p < 0.05$ were obtained

3. Results and Discussion

3.1. Iodine Content of Sugar Kelp

Fresh Norwegian sugar kelp harvested in April 2018 contained 4605 ± 274 mg iodine·kg^{-1}·dw^{-1}, which is comparable to other European cultivated sugar kelp (3460–6568 mg·kg^{-1}·dw^{-1}) [6,7,29,30]. The process of freeze-thawing sugar kelp did not decrease the iodine content significantly (one-way ANOVA; F = 117, df = 15, $p < 0.001$) (Table 1). However, water blanching decreased the iodine content significantly for all blanching treatments except when treated at 30 °C for 2 s. All blanching treatments, except 30 °C below 120 s and 45 °C at 2 s, sufficiently reduced the iodine content below the maximum level of 2000 mg·kg^{-1}·dw^{-1} as recommended by ANSES (2018) [12] in seaweed products. The iodine content in the blanched sugar kelp approached a constant level for various treatments with an average content of 328 ± 19 mg·kg^{-1}·dw^{-1} (Figure 1). Similarly, Stévant et al., (2018) [7] also reported that a constant level was achieved when subjecting *S. latissima* to warm water at 32 °C for 1 h. The most efficient treatment in this present study reduced the iodine content to 12% relative to the initial iodine content in fresh sugar kelp.

For a better perspective of the iodine content and safe intake of sugar kelp, the recommended intake (RI) and upper intake levels (UL) for adults are used [9–11]. If considering the only dietary source of iodine for an adult was from sugar kelp, then to reach the RI and UL 0.35 g or 1.4 g of fresh non-processed sugar kelp should be consumed, respectively. In the case of blanched sugar kelp then 9.2 or 37 g of sugar kelp could be consumed for RI and UL, respectively. A risk assessment considering other sources of iodine in a daily diet should be taken into consideration when evaluating the potential risk of sugar kelp consumption, but overall, this study proves that a reduction of iodine in sugar kelp can be obtained.

The iodine content reached a constant level at 120 s for the treatments at 45 °C and 60 °C, thus the treatments at those temperatures with a longer processing time (300 s) did not undergo further chemical analysis. The 30 °C treatments and the freeze-thawed treatments were also not further investigated, as the treatments did not reduce the iodine content as satisfactorily as the others.

Table 1. Iodine content in fresh, freeze-thawed and water-blanched *Saccharina latissima* expressed in mg·kg^{-1}·dw^{-1}. Results are mean ± standard deviation (n = 3).

Time	Temperature/Treatment	Iodine (mg·kg^{-1}·dw^{-1})
N/A	Fresh	4605 ± 274 ^{ab}
N/A	Freeze-thawed	4057 ± 419 ^{b}
2 s	30 °C	5157 ± 201 ^{a}
	45 °C	2873 ± 627 ^{c}
	60 °C	1198 ± 146 ^{d}
	80 °C	711 ± 151 ^{de}
30 s	45 °C	667 ± 120 ^{de*}
	60 °C	472 ± 121 ^{de}
	80 °C	343 ± 41 ^{e}
120 s	30 °C	2973 ± 523 ^{c}
	45 °C	346 ± 35 ^{e}
	60 °C	334 ± 55 ^{e}
	80 °C	293 ± 90 ^{e}
300 s	30 °C	1014 ± 349 ^{de}
	45 °C	388 ± 23 ^{de}
	60 °C	321 ± 68 ^{e}

N/A designates not applicable. (*) indicates two replicates (n = 2). Letters (a–e) denote significant differences between treatments by one-way ANOVA and Tukey's post-hoc test.

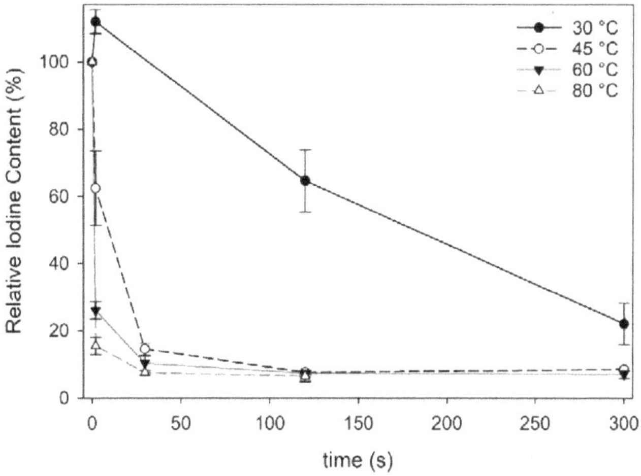

Figure 1. Iodine content in water blanched *Saccharina latissima* relative to fresh *S latissima* expressed in %. Each data point represent the mean iodine content with standard deviations (n = 3).

3.2. Proximate Composition

The proximate composition (n = 3) for the selected treatments can be found in Table 2. Ash content of fresh sugar kelp was 44.51 ± 0.86% dw. The content of ash varied significantly for all water-blanched samples (one-way ANOVA; F = 79, df = 10, p < 0.001). The ash content for water-blanched samples was between 9.1 ± 1.6% dw and 26.3 ± 1.5% dw. Protein content in fresh sugar kelp was 7.9 ± 2.5% dw, and in the blanched samples it was 9.8 ± 3.0% dw to 15.3 ± 2.6% dw. No significant differences were found for protein content between any of the samples (one-way ANOVA; F = 2.4, df = 10, p = 0.064). The lipid content was 5.8 ± 2.6% dw in fresh *S. latissima* and varied for the blanched samples between 6.9 ± 0.8% dw to 10.2 ± 0.6% dw with no significant differences (one-way ANOVA; F = 1.8, df = 10, p = 0.132). Carbohydrates were calculated from the other proximates. As the ash content showed significant difference between treatments, the carbohydrates also showed significant variations (one-way ANOVA; F = 14, df = 10, p < 0.001).

Table 2. Proximate composition of fresh and water-blanched *Saccharina latissima.* Data are expressed as means ± SD and represent three process replications (n = 3). Water is given in % ww, whereas ash, protein (total amino acids), fat, and calculated carbohydrates are given in % dw.

Component	Fresh	45 °C			60 °C			80 °C		
		2 s	30 s	120 s	2 s	30 s	120 s	2 s	30 s	120 s
Water	90.68 ± 0.30 ^{a}	93.42 ± 0.77 ^{b}	94.79 ± 0.47 ^{c}	95.70 ± 0.20 ^{c}	94.49 ± 0.46 ^{bc}	95.45 ± 0.20 ^{c}	95.44 ± 0.28 ^{c}	95.23 ± 0.20 ^{c}	95.36 ± 0.03 ^{*c}	95.64 ± 0.14 ^{c}
Ash	44.51 ± 0.86 ^{a}	26.3 ± 1.5 ^{b}	18.4 ± 1.7 ^{cd}	10.8 ± 2.5 ^{ef}	20.5 ± 3.2 ^{bc}	12.3 ± 2.8 ^{def}	9.1 ± 1.6 ^{f}	17.2 ± 1.7 ^{cde}	11.7 ± 1.2 ^{*def}	11.2 ± 1.4 ^{def}
Protein	7.9 ± 2.5 ^{a}	11.8 ± 2.4 ^{a}	10.5 ± 1.4 ^{*a}	12.3 ± 1.0 ^{a}	10.2 ± 3.0^{a}	9.8 ± 3.0 ^{a}	13.6 ± 1.8 ^{a}	12.6 ± 2.3 ^{a}	13.6 ± 2.3 ^{a}	15.3 ± 2.6 ^{a}
Fat	5.8 ± 2.6 ^{a}	6.9 ± 0.8 ^{a}	7.9 **	10.2 ± 0.6 ^{*a}	9.1 ± 1.5^{a}	8.6 ± 4.0 ^{*a}	9.0 ± 1.7 ^{a}	9.7 ± 0.7 ^{*a}	9.1 ± 1.5 ^{a}	8.7 ± 1.3 ^{*a}
Carbohydrates	41.8 ± 4.7 ^{a}	55.0 ± 0.3 ^{abc}	65.2 **	68.7 ± 3.9 ^{cd}	60.1 ± 5.0^{bcd}	65.3 ± 4.6 ^{*cd}	68.3 ± 1.4 ^{cd}	60.2 ± 1.8 ^{*cd}	64.9 ± 0.6 ^{*cd}	63.7 ± 1.4 ^{*cd}

(*) Included only duplicates (n = 2), (**) indicated one replicate (n = 1). Means with different letters (a–f) within each row are significantly different ($p \leq 0.05$).

3.3. Retention of Nutrients

The proximate composition is given for samples after each individual treatment, but does not take the potential loss of biomass into consideration due to processing. From the proximate composition, it cannot be concluded that there is a loss or gain due to treatment, therefore the true retention factors were calculated relative to fresh sugar kelp (Table 3). In Figure 2 the concentrations of each individual proximate and the true retention factors are seen. This together defines the mass balances of the treatments, which can indicate if there is a loss of each individual proximate. The total loss of ash, protein, lipid, and carbohydrate all together (total proximate) for each individual treatment are given relative to the fresh *S. latissima*. Fresh sugar kelp had a total proximate composition of 9.3 g 100 g^{-1} ww and the retained total amount of each treatment varied from 4.3 to 6.6 g 100 g^{-1} ww. The treatment that had the least loss (29%) of proximate was 45 °C at 2 s, whereas the others had a loss that ranged from 41% to 54%.

Table 3. True retention factors post processing relative to the fresh sugar kelp. The retention factors are presented in means ± SD (n = 3).

Component			45 °C			60 °C			80 °C
	2 s	30 s	120 s	2 s	30 s	120 s	2 s	30 s	120 s
Water	0.87 ± 0.03 ^{a}	0.99 ± 0.06 ^{a}	0.93 ± 0.13 ^{a}	0.83 ± 0.12 ^{a}	0.83 ± 0.04 ^{a}	0.87 ± 0.03 ^{a}	0.74 ± 0.05 ^{a}	0.81 ± 0.04 ^{a}	0.86 ± 0.13 ^{a}
Ash	0.39 ± 0.07 ^{b}	0.22 ± 0.02 ^{c}	0.09 ± 0.02 ^{d}	0.013 ± 0.005 ^{d}	0.005 ± 0.001 ^{d}	0.004 ± 0.000 ^{d}	0.007 ± 0.001 ^{d}	0.005 ± 0.001 ^{d}	0.004 ± 0.001 ^{d}
Protein	0.89 ± 0.22 ^{ab}	0.69 ± 0.05 ^{ab}	0.63 ± 0.11 ^{ab}	0.65 ± 0.28 ^{ab}	0.48 ± 0.16 ^{b}	0.70 ± 0.11 ^{ab}	0.57 ± 0.07 ^{ab}	0.67 ± 0.11 ^{ab}	0.73 ± 0.09 ^{ab}
Fat	0.73 ± 0.11 ^{a}	0.74 ± 0.06 ^{a}	0.69 ± 0.07 ^{a}	0.79 ± 0.23 ^{a}	0.60 ± 0.26 ^{a}	0.79 ± 0.23 ^{a}	0.60 ± 0.26 ^{a}	0.65 ± 0.14 ^{a}	0.62 ± 0.03 ^{a}
Carbohydrates	0.72 ± 0.12 ^{ab}	0.79 ± 0.10 ^{ab}	0.65 ± 0.08 ^{b}	0.89 ± 0.16 ^{ab}	0.74 ± 0.01 ^{ab}	0.74 ± 0.01 ^{ab}	0.66 ± 0.08 ^{b}	0.71 ± 0.04 ^{ab}	0.69 ± 0.10 ^{ab}

Means with different letters (a–f) within each row are significantly different ($p \leq 0.05$). The factors are relative to the fresh sugar kelp, which had a factor of 1.0 and statistical letter (a). Statistical descriptions: water (one-way ANOVA; F = 3.6, df = 10, p = 0.006), ash (one-way ANOVA; F = 297, df = 10, p < 0.001), protein (one-way ANOVA; F = 2.2, df = 10, p = 0.060), fat (one-way ANOVA; F = 1.3, df = 10, p = 0.287), and carbohydrates (one-way ANOVA; F = 2.7, df = 10, p = 0.025).

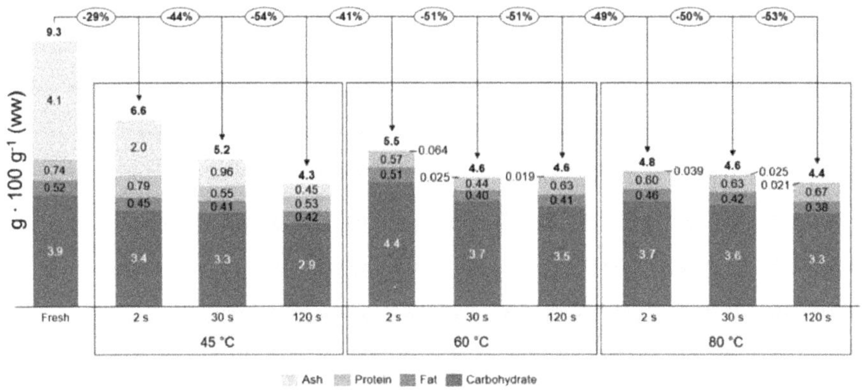

Figure 2. Mass balances for the proximate composition relative to the fresh sugar kelp for each blanching treatment. The percentages in circles describe the total loss of the proximate composition (excluding water). The concentration of each proximate is described by the bar diagram and the bold number above the bars are the total proximate composition relative to wet weight.

To the best of the authors' knowledge, no earlier study has explored the potential loss of proximate composition in *S. latissima*, due to water blanching. Therefore, comparisons to peer-reviewed studies on vegetables were performed. Water blanching (60, 150 and 180 s) of bell peppers, peas, and potatoes led to a significant protein loss of between 8% and 24% [31,32]. In this current study a significant difference in protein was only found for 60 °C at 30 s with a retention of 0.48 ± 0.16 (Tukey's post-hoc test; $p = 0.047$). In all other cases, no significant differences were found due to high standard deviations (Tukey's post-hoc test; $p > 0.154$). The true retention factor of protein was on average 0.67, meaning a total protein loss of 33%, which was higher than that found for vegetables. The high standard deviations are most likely due to the method of treating the product after blanching. The sugar kelp surface consists of mucus and seawater, which was probably interfering with the blanching water during treatment. If the mucus and seawater were washed away during blanching it would be replaced by blanching water on the surface. By shaking the sugar kelp consistently, it was expected that the blanching water would be removed, but some of the blanching water would stay on the product surface and interfere when weighing the samples, creating high standard deviations. The true retention factors and the mass balances indicated a significant loss for the ash content (one-way ANOVA; F = 297, df = 10, $p < 0.001$). *Saccharina latissima* is rich in minerals and trace elements such as Na, K, Mg, and Fe [33]. This significant loss of ash is probably not only due to the loss of iodine, but also other minerals and trace elements. If these minerals are located on the surface of the sugar kelp, they could dissolve into the water when blanched. Moreover, the relatively high processing temperature and low salinity of the blanching water could create a shock to the cells, leading to cell bursts and protein and minerals leaking from the cells.

3.4. Amino Acid Composition

The amino acid composition was quantified for selected water blanching treatments and the fresh sugar kelp sample. Two of the amino acids (glutamic acid and alanine) had a significant loss due to treatments when compared to the fresh sugar kelp (Figure 3). Fresh sugar kelp contained significantly higher amounts of glutamic acid (173 mg·g^{-1} protein; one-way ANOVA; F = 6.4, df = 10, $p < 0.001$) when compared to the treated samples, although not when compared to the 45 °C at 2 and 30 s (Tukey's post-hoc test; $p > 0.395$). This meant that the treatments with higher temperatures and process times had a significant loss of glutamic acid. On average, the significantly different samples had an average content of 146 mg glutamic acid g^{-1} protein. For alanine, the fresh sample was significantly different to the treated samples, meaning that there was a significant loss due to processing (one-way ANOVA; F = 52, df = 10, $p < 0.001$). The content in fresh sugar kelp was 178 mg alanine·g^{-1} protein and the average of the treated samples were 128 mg alanine·g^{-1} protein. The entire amino acid profile can be found in the data repository.

No significant changes were found for the essential amino acids. The essential amino acid to total amino acid ratio (EAA ratio) increased, since there was a significant loss of the non-essential alanine and glutamic acid. Fresh sugar kelp had an EAA ratio of 42.01 ± 0.59% EAA, and this was comparable to studies from Denmark and the Faroe Islands [2,34]. Whereas, the blanched samples on average had a ratio of 48.0 ± 1.2%, and were significantly different from the fresh sample (one-way ANOVA; F = 7.9, df = 10, $p < 0.001$).

The limiting amino acid for all samples was histidine, which is also seen in other studies [2,34]. The amino acid score (not considering digestion) was on average above 100% (108 ± 12%), with no significant differences between neither sample (one-way ANOVA; F = 1.1, df = 10, $p = 0.388$). Summing up, the blanching treatment did not compromise the amino acid quality but actually increased it as two non-essential amino acids had a significant loss.

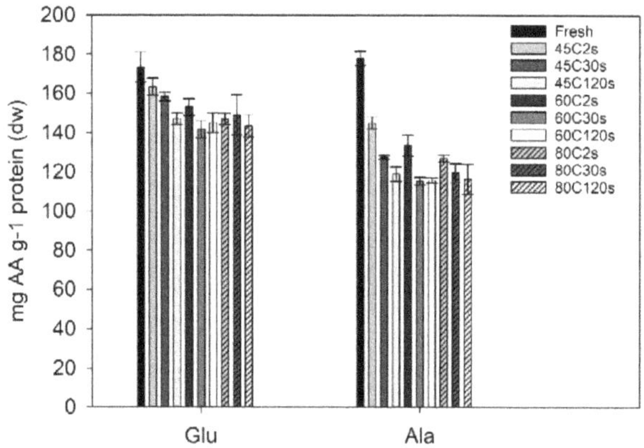

Figure 3. Glutamic acid and alanine in *Saccharina latissima* after different treatments given in mg amino acid (AA) g^{-1} of protein. Error bars represent the standard deviation (n = 3). A one-way ANOVA indicated a significant difference between the fresh sugar kelp sample compared to the treated sugar kelp for both glutamic acid (Glu) and alanine (Ala).

3.5. Fatty Acid Composition

The fatty acid (FA) profile was quantified by direct methylation and given in % FAME for fresh sugar kelp and the samples blanched at 45 °C and 60 °C for 30 s and 300 s, respectively. The complete FA profile (% FAME) can be found in the data repository. The quality of FAs can be explained by the content of the individual fatty acids, which the human body cannot synthesize—α-linolenic acid (ALA) and linoleic acid (LA), but also the two fatty acids docosahexaenoic acid (DHA) and eicosapentaenoic acid (EPA). Moreover, the total amount of polyunsaturated fatty acids (PUFA) and omega-3 fatty acids (n-3) as well as the ratio of n-3 FA to omega-6 fatty acids (n-6) indicate the quality of the lipid fraction. The 60 °C 300 s blanched samples presented a higher proportion of EPA, ALA, PUFA, and n-3, and a higher n-3/n-6 ratio compared to the fresh and 45 °C 30 s blanched samples (Table 4). The increased proportion results from the reduction of other fatty acids, namely unsaturated and monounsaturated, during the processing. Overall, this results in a biomass with an improved profile of health-beneficial fatty acids. No significant difference was found for LA, while DHA, which was present in the sugar kelp in very low amounts, seem to be degraded during processing.

Table 4. Fatty acid composition of fresh and blanched sugar kelp expressed in % FAME. Data are expressed as means ± SD and represents three process replications (n = 3). The fatty acids are given in % FAME although the ratio (n-3/n-6) is without unit.

Fatty Acids	Fresh	45 °C	60 °C
		30 s	300 s
18:2 (n-6) (LA)	4.96 ± 0.12 ^{ab}	5.50 ± 0.23 ^{a}	4.87 ± 0.16 ^{b}
18:3 (n-3) (ALA)	15.2 ± 1.5 ^{a}	18.1 ± 2.1 ^{ab}	22.63 ± 0.45 ^{b}
20:5 (n-3) (EPA)	12.18 ± 0.82 ^{a}	13.2 ± 1.0 ^{a}	17.38 ± 0.16 ^{b}
22:6 (n-3) (DHA)	0.36 ± 0.02 ^{a}	0.15 ± 0.08 ^{b}	0.00 ± 0.00 ^{c}
n-3	29.0 ± 2.4 ^{a}	32.2 ± 3.0 ^{a}	41.20 ± 0.71 ^{b}
n-6	22.51 ± 0.91 ^{a}	25.37 ± 0.61 ^{b}	26.00 ± 0.03 ^{b}
n-3/n-6	1.29 ± 0.08 ^{a}	1.27 ± 0.09 ^{a}	1.59 ± 0.03 ^{b}
PUFA	51.5 ± 3.1 ^{a}	57.6 ± 3.6 ^{a}	67.19 ± 0.67 ^{b}

(a–e) denote significant difference between sample treatments. From the top linoleicacid (LA), α-linolenic acid (ALA), eicosapentaenoic acid (EPA), docosahexaenoic acid (DHA), total omega-3 fatty acids (n-3), total omega-6 fatty acids (n-6), the omega-3 to omega-6 fatty acids ratio (n-3/n-6), and total polyunsaturated fatty acids (PUFA).

3.6. Antioxidant Activity and Total Phenolic Content

Processing had a significant effect on the amount of methanolic extract obtained (one-way PERMANOVA; F = 20.5, df = 3, $p = 0.002$), with significantly higher amounts extracted from the fresh samples compared to the water-blanched samples (7.2–11.1% dw; Figure 4). This demonstrates that blanching sugar kelp will result in a significant amount of compounds being transferred to the water phase or degraded during processing.

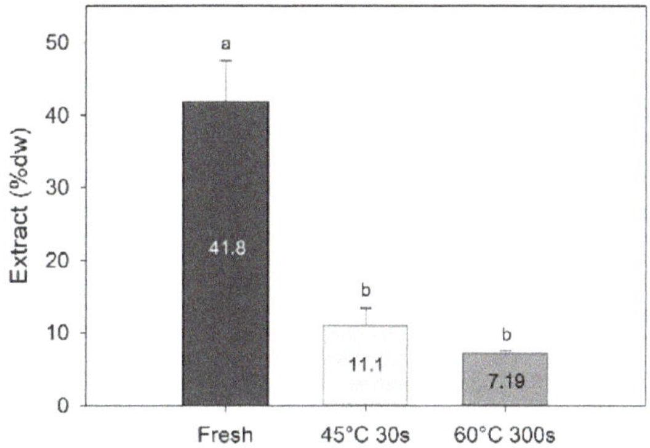

Figure 4. Amount of methanolic extracts (% dw) of sugar kelp for fresh and two different blanching treatments. Different letters represent a significant difference ($p < 0.05$) between treatments. Data are mean ± SD; $n = 3$.

Blanching had a significant effect on total phenolic content (TPC) (one-way PERMANOVA; F = 26.0, df = 3, $p = 0.0011$; Figure 5). TPC was higher in the 60 °C 300 s treated sample ($p = 0.013$) compared to the fresh sample. On the other hand, there was no significant difference in TPC between fresh and 45 °C 30 s blanched samples ($p = 0.07$). TPC found in the present study for fresh is within the range of the values reported for *S. latissima* harvested at different seasons (0.84–2.41 mg·GAE/g sugar kelp [5]). On the other hand, TPC values obtained for the blanched samples are above those reported in the same study.

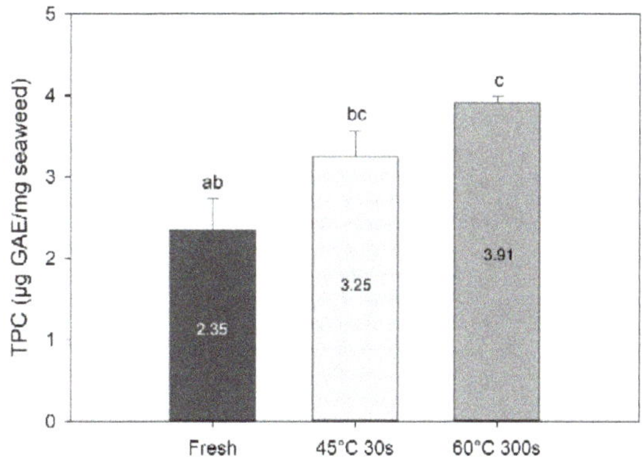

Figure 5. Total phenolic content of fresh and selected blanched samples of *Saccharina latissima* expressed in gallic acid equivalents per mg of freeze-dried samples. Data are mean ± SD; $n = 3$. Different letters represent a significant difference between treatments ($p < 0.05$).

TPC results expressed in gallic acid equivalents per mg of extract revealed an even greater effect of processing on the TPC. Blanching significantly increased the content of TPC as compared to the fresh sugar kelp (one-way PERMANOVA; F = 392, df = 3, $p < 0.01$, Figure 6). The highest TPC was found in the 60 °C 300 s blanched samples ($p < 0.013$), followed by the 45 °C 30 s blanched samples, and then fresh samples ($p = 0.96$). These results suggest that the extraction of other compounds during water blanching may have resulted in concentration of phenolic compounds in the processed sugar kelp.

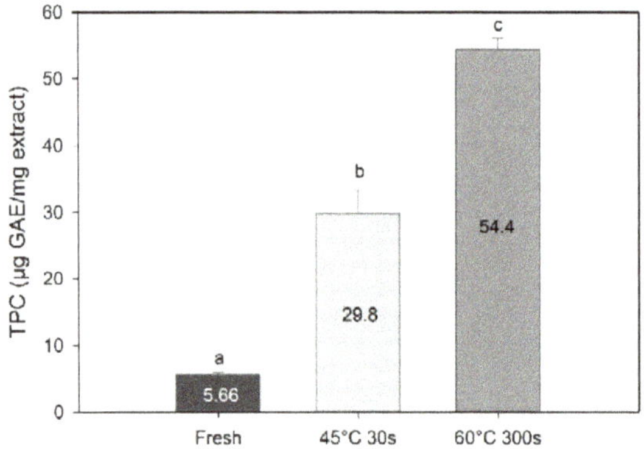

Figure 6. Total phenolic content expressed in gallic acid equivalents per mg of methanolic extract of *Saccharina latissima* from fresh and two types of blanching. Data are mean ± SD; $n = 3$. Different letters represent a significant difference ($p < 0.05$) between treatments.

DPPH radical scavenging activity revealed a concentration dependency and increased with increasing concentrations of algal extract (data not shown). Processing increased the radical scavenging activity significantly (F = 13.5, df = 3, $p = 0.0053$, Figure 7). These results suggest that compounds with high radical scavenging activity are retained and up-concentrated in the sugar kelp during water blanching. This correlates well with the current results; TPC have been identified as a major component contributing to radical scavenging activity of seaweed [5,35–38].

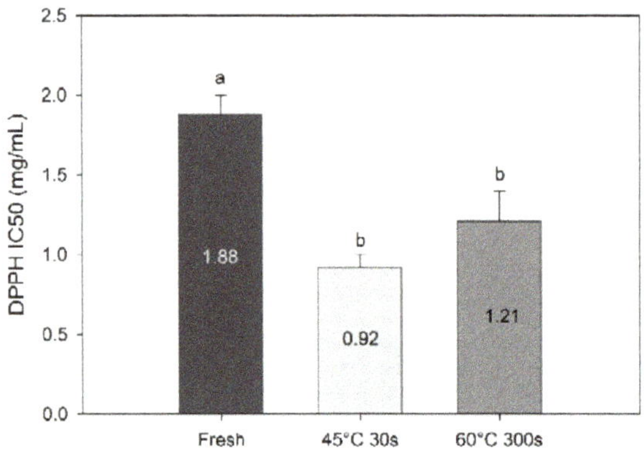

Figure 7. DPPH radical scavenging activity (IC50; mg/mL) of methanolic extracts of fresh and blanched *Saccharina latissima*. Different letters represent a significant difference ($p < 0.05$) between treatments. Data are mean ± SD; $n = 3$.

4. Conclusions

This study showed that water blanching is a promising approach for reducing the iodine content in Norwegian-cultivated *Saccharina latissima*. Up to 88% reduction was obtained by blanching at optimized conditions (≥45 °C and ≥30 s. Considering the recommended intake and upper intake level reported by the Nordic Nutrition Recommendations (2012)). If sugar kelp was the only source of dietary iodine, a maximum 9.2 g or 37 g, respectively of blanched sugar kelp can be consumed daily to avoid exceeding these recommendations. However, freeze-thawing did not decrease the iodine content of sugar kelp. These are important findings for the food-producing industry that is using seaweed as a raw material and is responsible for consumer safety. In terms of processing effects on other nutritionally valuable compounds, the treatment that had the least loss (29%) of total proximate composition was 45 °C at 2 s, whereas the other treatments had a loss that ranged from 41% to 54%. More specifically, a significant loss of ash occurred, which is comparable with the degree of loss of iodine together with other minerals. Water blanching also caused a significant loss of two amino acids (glutamic acid and alanine), which led to a higher EAA/AA ratio. Moreover, water blanching resulted in biomass with an improved composition of health beneficial compounds, namely PUFA and phenolic compounds, and antioxidant activity.

In perspective, other valuable compounds with antioxidant activity found in sugar kelp such as the pigment fucoxanthin and carbohydrates could have been interesting to study. Moreover, the change in texture, color, and taste (e.g., umami) due to blanching is also interesting and worth further study.

Author Contributions: Conceptualization, J.F., M.S. and C.W.N.; chemical analysis, C.W.N. and G.S.M.; resources, J.F.; writing—original draft preparation, C.W.N. and G.S.M.; writing—review and editing, C.W.N., G.S.M., M.S., J.J.S., S.L.H. and T.R.; supervision, J.J.S., S.L.H. and T.R. All authors have read and agreed to the published version of the manuscript.

Funding: This research received no external funding.

Conflicts of Interest: The authors declare no conflict of interest.

References

1. Holdt, S.L.; Kraan, S. Bioactive compounds in seaweed: Functional food applications and legislation. *J. Appl. Phycol.* **2011**, *23*, 543–597. [CrossRef]
2. Bak, U.G.; Nielsen, C.W.; Marinho, G.S.; Gregersen, Ó.; Jónsdóttir, R.; Holdt, S.L. The seasonal variation in nitrogen, amino acid, protein and nitrogen-to-protein conversion factors of commercially cultivated Faroese Saccharina latissima. *Algal Res.* **2019**, *42*. [CrossRef]
3. Nielsen, M.M.; Manns, D.; D'Este, M.; Krause-Jensen, D.; Rasmussen, M.B.; Larsen, M.M.; Alvarado-Morales, M.; Angelidaki, I.; Bruhn, A. Variation in biochemical composition of Saccharina latissima and Laminaria digitata along an estuarine salinity gradient in inner Danish waters. *Algal Res.* **2016**, *13*, 235–245. [CrossRef]
4. Biancarosa, I.; Belghit, I.; Bruckner, C.G.; Liland, N.S.; Waagbø, R.; Amlund, H.; Heesch, S.; Lock, E.J. Chemical characterization of 21 species of marine macroalgae common in Norwegian waters: Benefits of and limitations to their potential use in food and feed. *J. Sci. Food Agric.* **2018**, *98*, 2035–2042. [CrossRef] [PubMed]
5. Marinho, G.S.; Sørensen, A.D.M.; Safafar, H.; Pedersen, A.H.; Holdt, S.L. Antioxidant content and activity of the seaweed Saccharina latissima: A seasonal perspective. *J. Appl. Phycol.* **2019**, *31*, 1343–1354. [CrossRef]
6. Schiener, P.; Black, K.D.; Stanley, M.S.; Green, D.H. The seasonal variation in the chemical composition of the kelp species Laminaria digitata, Laminaria hyperborea, Saccharina latissima and Alaria esculenta. *J. Appl. Phycol.* **2014**, *27*, 363–373. [CrossRef]
7. Stévant, P.; Marfaing, H.; Duinker, A.; Fleurence, J.; Rustad, T.; Sandbakken, I.; Chapman, A. Biomass soaking treatments to reduce potentially undesirable compounds in the edible seaweeds sugar kelp (Saccharina latissima) and winged kelp (Alaria esculenta) and health risk estimation for human consumption. *J. Appl. Phycol.* **2018**, 1–14. [CrossRef]
8. Danish Food Data version 4. Available online: https://frida.fooddata.dk (accessed on Dec 31, 2019).
9. Nordic Council of Ministers. *Nordic Nutrition Recommendations 2012 - Integrating Nutrition and Physical Activity*, 5th ed.; Nordic Council of Ministers: Copenhagen, Denmark, 2012; ISBN 9789289326704.
10. *Scientific Commitee on Food Tolerable Upper Intake Levels for Vitamins and Minerals*; European Food Safety Authorities: Parma, Italy, 2006; ISBN 9291990140.
11. *EFSA Overview on Tolerable Upper Intake Levels as derived by the Scientific Committee on Food (SCF) and the EFSA Panel on Dietetic Products, Nutrition and Allergies (NDA)*; European Food Safety Authorities: Parma, Italy, 2018.

12. *ANSES Opinion of the French Agency for Food, Environmental and Occupational Health and Safety: On the Risk of Excess Iodine Intake from the Consumption of Seaweed in Foodstuffs*; Request No 2017-SA-0086; ANSES (French Agency for Food, Environmental and Occupational Health & Safety: Maisons-Alfort, France, 2018.
13. Council directive (EC) 178/2002 laying down the general principles and requirements of food law, establishing the European Food Safety Authority and laying down procedures in matters of food safety. *Off. J.* **2002**, *L31*, 1–24.
14. Lüning, K.; Mortensen, L. European aquaculture of sugar kelp (Saccharina latissima) for food industries: Iodine content and epiphytic animals as major problems. *Bot. Mar.* **2015**, *58*, 449–455. [CrossRef]
15. *AOAC Method 950.46 Official Methods of Analysis*; Association of Official Analytical Chemists: Washington, DC, USA, 1990.
16. *AOAC Method 938.08 Official Methods of Analysis*; Association of Official Analytical Chemists: Washington, DC, USA, 1990.
17. *EN 17050:2017 Animal Feeding Stuffs: Methods of Sampling and Analysis—Determination of Iodine in Animal Feed by ICP-MS*; European Commitee for Standardization (CEN): Brussels, Belgium, 2017.
18. Angell, A.R.; Mata, L.; de Nys, R.; Paul, N.A. The protein content of seaweeds: A universal nitrogen-to-protein conversion factor of five. *J. Appl. Phycol.* **2016**, *28*, 511–524. [CrossRef]
19. *FAO Food Energy—Methods of Analysis and Conversion Factors*; FAO: Rome, Italy, 2003.
20. Diniz, G.S.; Barbarino, E.; Oiano-Neto, J.; Pacheco, S.; Lourenco, S.O. Gross Chemical Profile and Calculation of Nitrogen-to-Protein Conversion Factors for Five Tropical Seaweeds. *Am. J. Plant Sci.* **2011**, *02*, 287–296. [CrossRef]
21. Bligh, E.G.; Dyer, W.J. A rapid method of total lipid extraction and purification. *Can. J. Biochem. Physiol.* **1959**, *37*, 911–917. [CrossRef] [PubMed]
22. Safafar, H.; Ljubic, A.; Møller, P.; Jacobsen, C. Two-Step Direct Transesterification as a Rapid Method for the Analysis of Fatty Acids in Microalgae Biomass. *Eur. J. Lipid Sci. Technol.* **2019**, *121*, 1–8. [CrossRef]
23. Firestone, D. *Official Methods and Recommended Practices of the AOCS*, 5th ed.; American Oil Chemists' Society: Urbana, IL, USA, 1998.
24. Safafar, H.; Van Wagenen, J.; Møller, P.; Jacobsen, C. Carotenoids, phenolic compounds and tocopherols contribute to the antioxidative properties of some microalgae species grown on industrial wastewater. *Mar. Drugs* **2015**, *13*, 7339–7356. [CrossRef] [PubMed]
25. Farvin, K.H.S.; Jacobsen, C. Phenolic compounds and antioxidant activities of selected species of seaweeds from Danish coast. *Food Chem.* **2013**, *138*, 1670–1681. [CrossRef]
26. Yang, J.; Guo, J.; Yuan, J. In vitro antioxidant properties of rutin. *LWT - Food Sci. Technol.* **2008**, *41*, 1060–1066. [CrossRef]
27. Murphy, E.W.; Criner, P.E.; Gray, B.C. Comparisons of Methods for Calculating Retentions of Nutrients in Cooked Foods. *J. Agric. Food Chem.* **1975**, *23*, 1153–1157. [CrossRef]
28. Anderson, M.J.; Gorley, R.N.; Clarke, K.R. *PERMANOVA+ for PRIMER: Guide to Software and Statistical Methods*; PRIMER-E: Plymouth, UK, 2008.
29. Nitschke, U.; Stengel, D.B. A new HPLC method for the detection of iodine applied to natural samples of edible seaweeds and commercial seaweed food products. *Food Chem.* **2015**, *172*, 326–334. [CrossRef]
30. Roleda, M.Y.; Skjermo, J.; Marfaing, H.; Jónsdóttir, R.; Rebours, C.; Gietl, A.; Stengel, D.B.; Nitschke, U. Iodine content in bulk biomass of wild-harvested and cultivated edible seaweeds: Inherent variations determine species-specific daily allowable consumption. *Food Chem.* **2018**, *254*, 333–339. [CrossRef]
31. Adams, J.B. Blanching of Vegetables. *Nutr. Food Sci.* **1981**, *81*, 11–13. [CrossRef]
32. Castro, S.M.; Saraiva, J.A.; Lopes-da-Silva, J.A.; Delgadillo, I.; Van Loey, A.; Smout, C.; Hendrickx, M. Effect of thermal blanching and of high pressure treatments on sweet green and red bell pepper fruits (Capsicum annuum L.). *Food Chem.* **2008**, *107*, 1436–1449. [CrossRef]
33. Circuncisão, A.R.; Catarino, M.D.; Cardoso, S.M.; Silva, A.M.S. Minerals from macroalgae origin: Health benefits and risks for consumers. *Mar. Drugs* **2018**, *16*, 400. [CrossRef] [PubMed]
34. Marinho, G.S.; Holdt, S.L.; Angelidaki, I. Seasonal variations in the amino acid profile and protein nutritional value of Saccharina latissima cultivated in a commercial IMTA system. *J. Appl. Phycol.* **2015**, *27*, 1991–2000. [CrossRef]
35. Cho, S.H.; Kang, S.E.; Cho, J.Y.; Kim, A.R.; Park, S.M.; Hong, Y.K.; Ahn, D.H. The antioxidant properties of brown seaweed (Sargassum siliquastrum) extracts. *J. Med. Food* **2007**, *10*, 479–485. [CrossRef]

36. Wang, T.; Jónsdóttir, R.; Ólafsdóttir, G. Total phenolic compounds, radical scavenging and metal chelation of extracts from Icelandic seaweeds. *Food Chem.* **2009**, *116*, 240–248. [CrossRef]
37. Wang, T.; Jónsdóttir, R.; Liu, H.; Gu, L.; Kristinsson, H.G.; Raghavan, S.; Ólafsdóttir, G. Antioxidant capacities of phlorotannins extracted from the brown algae Fucus vesiculosus. *J. Agric. Food Chem.* **2012**, *60*, 5874–5883. [CrossRef]
38. Hermund, D.B.; Yeşiltaş, B.; Honold, P.; Jónsdóttir, R.; Kristinsson, H.G.; Jacobsen, C. Characterisation and antioxidant evaluation of Icelandic F. vesiculosus extracts in vitro and in fish-oil-enriched milk and mayonnaise. *J. Funct. Foods* **2015**, *19*, 828–841. [CrossRef]

foods

MDPI

Article

Development of Fish Oil-Loaded Microcapsules Containing Whey Protein Hydrolysate as Film-Forming Material for Fortification of Low-Fat Mayonnaise

Nor E. Rahmani-Manglano [1], Irene González-Sánchez [1], Pedro J. García-Moreno [1,*], F. Javier Espejo-Carpio [1], Charlotte Jacobsen [2] and Emilia M. Guadix [1]

1 Department of Chemical Engineering, University of Granada, 18071 Granada, Spain; norelenarm@ugr.es (N.E.R.-M.); irenegs@correo.ugr.es (I.G.-S.); fjespejo@ugr.es (F.J.E.-C.); eguadix@ugr.es (E.M.G.)

2 Division of Food Technology, National Food Institute, Technical University of Denmark, 2800 Lyngby, Denmark; chja@food.dtu.dk

* Correspondence: pjgarcia@ugr.es; Tel.: +34-958-2410-10 (ext. 20624)

Received: 31 March 2020; Accepted: 21 April 2020; Published: 30 April 2020

Abstract: The influence of the carbohydrate-based wall matrix (glucose syrup, GS, and maltodextrin, MD21) and the storage temperature (4 °C or 25 °C) on the oxidative stability of microencapsulated fish oil was studied. The microcapsules (ca. 13 wt% oil load) were produced by spray-drying emulsions stabilized with whey protein hydrolysate (WPH), achieving high encapsulation efficiencies (>97%). Both encapsulating materials showed an increase in the oxidation rate with the storage temperature. The GS-based microcapsules presented the highest oxidative stability regardless of the storage temperature with a peroxide value (PV) of 3.49 ± 0.25 meq O_2/kg oil and a content of 1-penten-3-ol of 48.06 ± 9.57 ng/g oil after six weeks of storage at 4 °C. Moreover, low-fat mayonnaise enriched with GS-based microcapsules loaded with fish oil and containing WPH as a film-forming material (M-GS) presented higher oxidative stability after one month of storage when compared to low-fat mayonnaise enriched with either a 5 wt% fish oil-in-water emulsion stabilized with WPH or neat fish oil. This was attributed to a higher protective effect of the carbohydrate wall once the microcapsules were incorporated into the mayonnaise matrix.

Keywords: omega-3; microencapsulation; spray-drying; whey protein; lipid oxidation; food fortification

1. Introduction

Food fortification with omega-3 polyunsaturated fatty acids (ω-3 PUFAs) has gained increased scientific and industrial interest in the last decades [1]. This is mainly due to the health benefits attributed especially to eicosapentaenoic (EPA; C20:5 n-3) and docosahexaenoic (DHA; C22:6 n-3) fatty acids [2,3]. However, the high number of bis-allylic hydrogens present in these ω-3 PUFAs make them highly prone to oxidation, leading to: (i) the loss of their nutritional properties and (ii) the appearance of odor/flavor-active and/or other potentially toxic compounds [4]. Hence, the development of efficient delivery systems which prevent ω-3 PUFAs oxidation prior (e.g., delivery system production); during (e.g., food processing) and after its incorporation into complex food matrices (e.g., food storage) is of great importance. In this regard, encapsulation technologies are of special interest for the food industry, since they allow the design of functional food systems overcoming the inherent drawbacks of ω-3 PUFAs rich oils (e.g., fish oil), such as low oxidative stability, low solubility and oily texture. Encapsulation consists of entrapping the core material (e.g., fish oil) within a homogeneous/heterogeneous matrix (e.g., encapsulating agent/s) to develop a physical barrier between the bioactive compound and the environment, thus preventing its degradation, easing its handling and/or controlling the bioactive release [5]. Among all the available encapsulation techniques (e.g., freeze-drying or coacervation), spray-dying is the most commonly used by the food industry to encapsulate bioactive ingredients [6]. Fish oil microencapsulation by spray-drying has been widely studied over the last decade, being the most commonly used encapsulating agents of carbohydrates, proteins and their combinations [7].

Nonetheless, additional stabilization techniques are required (e.g., addition of antioxidants to the formulation), since it has been demonstrated that the emulsification process and subsequent drying using air at high temperatures (170–200 °C) result in the initial oxidation of the oil [8,9]. Bioactive compounds such as protein hydrolysates or peptides, which exhibit both emulsifying and antioxidants properties, are a promising alternative to conventional antioxidants (e.g., tocopherols). In heterogeneous systems such as fish oil-in-water emulsions (which is the feed to the spray-drier), the location of the antioxidants determines their antioxidant activity [10,11]. The same is also the case for the resulting microcapsule. Antioxidants located at the oil/water interface (in emulsions) or at the oil/encapsulating agent interface (in spray-dried microcapsules) will be preferred to inhibit lipid oxidation. This is because these interfaces are the place where oxidation is started by the contact of prooxidants (e.g., trace of metals and oxygen) and the oil.

Whey protein hydrolysates (WPH) have been reported to exhibit antioxidant activity (e.g., radical scavenging activity, metal chelating and reducing power) [12,13]. Moreover, due their high emulsifying activity, WPH have also been used to produce fish oil microcapsules in combination with maltodextrin (DE 16.5–19.5) [14,15]. Nevertheless, the protective effect of WPH on the reduction of lipid oxidation in the microcapsules was not investigated in any of the latter studies. More recently, Tamm et al. [16] produced fish oil-loaded microencapsulates using glucose syrup as the wall material in the presence of unmodified or hydrolyzed β-lactoglobulin (β-LG). The unmodified and hydrolyzed proteins were used as a film-forming material around the oil droplets. The authors found that β-LG hydrolysates enhanced the oxidative stability of microencapsulated fish oil when compared to the unmodified protein. This fact was attributed to an improved accessibility of the resulting amino acid residues with antioxidant properties after enzymatic hydrolysis. Thus, the results reported by Tamm et al. [17] showed a promising potential for the use of spray-dried microcapsules loaded with fish oil and obtained with WPH as a delivery system of ω-3 PUFAs. However, the enrichment of food products with ω-3 microencapsulates having WPH as a film-forming material, and its effect on the physicochemical properties of the fortified product remains to be evaluated.

In this regard, the aims of the present study were to investigate: (i) the effect of the carbohydrate-based wall material (glucose syrup or maltodextrin) and (ii) the storage temperature (4 °C or 25 °C) on the oxidative stability of spray-dried microcapsules loaded with fish oil and produced using WPH as a film-forming material. In addition, the feasibility of using spray-dried microcapsules containing WPH for fortifying a low-fat mayonnaise (40 wt% of total oil) with ω-3 PUFAs was assayed. For that purpose, the physical and oxidative stabilities of low-fat mayonnaise enriched either with microcapsules loaded with fish oil, a fish oil-in-water emulsion stabilized with WPH or neat fish oil were investigated during one month of storage.

2. Materials and Methods

2.1. Materials

Whey protein (34.6 wt% protein content) and maltodextrin (DE 21) were kindly provided by Abbott (Granada, Spain), while the glucose syrup (DE38, C*Dry 1934) was kindly donated by Cargill Germany GmbH (Krefeld, Germany). Alcalase 2.4 L was purchased from Novozymes (Bagsvaerd, Denmark). Refined fish oil (Omega Oil 1812 TG Gold) was acquired from BASF Personal Care and Nutrition GmbH (Illertissen, Germany) and stored at −80 °C until use. The fatty acid composition of the fish oil was determined by gas chromatography (GC), as described in [17], and it was as follows (major fatty acids in %, *w/w*): 7.0% myristic acid (C14:0), 16.7% palmitic acid (C16:0), 8.8% palmitoleic acid (C16:1n-7), 4.1% stearic acid (C18:0), 8.2% oleic acid (C18:1n-9), 19.3% eicosapentaenoic acid (C20:5n-3) and 16.1% docosahexaenoic acid (C22:6n-3). Peroxide value (PV) of the fish oil was measured as described in Section 2.5.4 and was 0.36 ± 0.03 meq O_2/kg oil. Tocopherol content (TC) of the fresh oil was measured as described in Section 2.5.4 and was: 427.0 ± 0.0 µg/g oil, 48.0 ± 1.4 µg/g oil, 1891.0 ± 12.7 µg/g oil and 644.3 ± 4.1 µg/g oil for alpha-, beta-, gamma- and delta-tocopherol, respectively. Refined sunflower oil (SFO), for the production of mayonnaise, was purchased from the local market. The specified fats compositions as given by the supplier were: 10.77 wt% saturated fats, 30.47 wt% monounsaturated fats and 58.76 wt% polyunsaturated fats. Peroxide value (PV) of the sunflower oil was measured as described in Section 2.7.2 and was of 3.5 ± 0.2 meq O_2/kg oil. Stabilizer Grinsted FF 1149 was kindly donated by DuPont (DuPont Nutrition Biosciences Aps, Haderslev, Denmark). The rest of the ingredients used in the production of the mayonnaises were purchased in the local market. The rest of reagents used for analysis were of analytical grade.

2.2. Enzymatic Hydrolysis of Whey Protein

Enzymatic hydrolysis of whey protein (WP) was carried out in an automatic titrator 718 Stat Titrino (Metrohm AG, Herisau, Switzerland) to a degree of hydrolysis 10% (DH 10) with alcalase. For this purpose, a solution containing 36 g of protein was prepared with distilled water to a final volume of 0.9 L. The process conditions were set to 50 °C and the pH to 8, and the enzyme-substrate ratio was fixed to 0.55 (*w/w*). The degree of hydrolysis was estimated with the pH-stat-method, as described by Camacho et al. [18]. The hydrolysis reaction took ca. 1.5 h, and the enzyme was deactivated at 100 °C for 5 min. The whey protein hydrolysate (WPH) solution was stored at −20 °C until further use.

2.3. Microencapsulation of Fish Oil by Spray-Drying

Previous to spray-drying, fish oil-in-water emulsions stabilized with WPH and containing one of the encapsulating agents were produced. The aqueous phase of the emulsions was prepared by dissolving the glucose syrup (GS) or maltodextrin (MD21) (28%, *w/w*) in the WPH solution and adding distilled water in order to have a protein content of 2 wt%. Then, the pH was adjusted to pH 8. A prehomogenization process was carried out for 2 min at 15,000 rpm using an Ultraturrax T-25 homogenizer (IKA, Staufen, Germany), while the oil (5%, *w/w*) was added during the first minute. The coarse emulsions were then homogenized in a high-pressure homogenizer (PandaPLUS 2000; GEA Niro Soavi, Lübeck, Germany) at a pressure range of 450/75 bar, applying 3 passes. The temperature during the emulsification process was kept under 32 °C. Subsequently, the emulsions were dried in a pilot plant scale spray-drier (Mobile Minor; Niro A/S, Copenhagen, Denmark) at 190/80 °C inlet/outlet temperature, respectively. The pressure of the pneumatic air activating the rotary atomizer was set to 4 bar, which implies a rotational speed of the atomizer of 22,000 rpm.

2.4. Oil Droplet Size Distribution (ODSD)

The oil droplet size distribution (ODSD) of both the parent emulsions (emulsions fed to the spray-drier) at day 0 and the reconstituted emulsions (emulsions resulting of redispersing the microcapsules in water) at days 0 (week 0), 14 (week 2), 28 (week 4) and 42 (week 6) were measured by laser diffraction in a Mastersizer 2000 (Malvern Instruments, Ltd., Worcestershire, UK), as described by García-Moreno et al. [19]. The emulsions were reconstituted by dissolving the powder in distilled water in order to achieve the same solids content as the original emulsion. Measurements were made in duplicate, and the results are given in 90th percentile.

2.5. Physicochemical Characterization of Microencapsulates

2.5.1. Moisture Content (MC) and Water Activity (a_w)

The moisture content of the microcapsules was determined using an infrared balance (AD-471A, Tokyo, Japan) where ca. 1 g of powder was heated at 105 °C for 90 min until a constant weight. The water activity was measured using a LabMASTER-aw (Novasina AG, Lachen, Switzerland) at 20 °C. The measurements were carried out in duplicates.

2.5.2. Encapsulation Efficiency (EE)

The encapsulation efficiency (EE) was determined by extracting the surface oil, as described by Danviriyakul et al. [20], with some modifications. In brief, 2.5 g of powder was weighed and mixed with 15 mL of hexane in a vortex mixer for 2 min and then centrifuged at 2720× *g* for 20 min. Five mL of supernatant were collected on a Pyrex tube, previously weighed, and the solvent was evaporated under a constant flow of nitrogen. After total solvent evaporation, in order to calculate the surface oil, the Pyrex was weighed again, and the oil concentration was adjusted to the initial volume of hexane added. The *EE* was calculated as follows:

$$EE, \% = \frac{A - B}{A} \times 100 \tag{1}$$

where A refers to the total theoretical amount of oil (g) and B to the non-encapsulated oil (g). Measurements were carried out in triplicate.

2.5.3. Morphology and Size

The morphology and size of the microcapsules were studied by means of scanning electron microscopy (SEM) using a FEI microscope (FEI Inspect, Hillsboro, OR, USA). For this purpose, a thin layer of microcapsules was placed on a carbon tape and sputter-coated with gold, 8 s, 40 mA using a Cressington 208HR Sputter-Coater (Cressington Scientific Instruments, Watford, England). The SEM images were taken in the range 830×–870× magnification with a 10-kV accelerating voltage. Then, they were analyzed using the ImageJ software (National Institute of Health). To determine the mean diameters of the microcapsules, more than 150 randomly selected microcapsules were measured.

2.5.4. Oxidative Stability

Spray-dried microcapsules loaded with fish oil were stored at 4 °C and 25 °C during 6 weeks in brown bottles (30 mL and 26-mm inner diameter). Each bottle contained 10 g of microcapsules. Samples were taken at days 0 (week 0), 7 (week 1), 14 (week 2), 21 (week 3), 28 (week 4), 35 (week 5) and 42 (week 6) and placed at −80 °C under a nitrogen atmosphere until the determination of the peroxide value (PV), tocopherol content (TC) and content of secondary volatile oxidation products (SVOP) was carried out.

Peroxide Value (PV)

Fish oil was extracted from the microcapsules, as described by Bligh and Dyer [21] using a reduced amount of chloroform/methanol (1:1, *w*/*w*) solvent. For the extraction, ca. 2 g of powder were weighed and then dissolved by adding 10 mL of distilled water. Peroxide value (PV) was then quantified on the lipid extracts using the colorimetric ferric-thiocyanate method at 500 nm, according to Shantha and Decker [22]. In brief, the extracted oil was diluted in chloroform/methanol (7:3, *v*/*v*) prior to the addition of iron-II-chloride and ammonium thiocyanate solutions. Then, the mixture was incubated for 5 min at room temperature. Measurements were carried out in duplicates. Results were expressed in meq O_2 per kg of oil.

Tocopherol Content (TC)

Tocopherol content of the microencapsulated oil was determined by HPLC (Agilent 1100 Series) according to the American Oil Chemists' Society (AOCS) official method [23]. In brief, about 2 g of the chloroform extract was evaporated under nitrogen and dissolved in 1-mL n-heptane, and from this, 0.8 mL were taken into separate vials before injection of an aliquot (40 µL) on a Spherisorb S5W column (250 × 4.6 mm) (Phase Separation Ltd., Deeside, UK). Elution was performed with an isocratic mixture of n-heptane/2-propanol (100:0.4, *v*/*v*) at a flow of 1 mL/min. Detection was done using a fluorescence detector with excitation at 290 nm and emission at 330 nm, according to the AOCS [23]. Measurements were performed in duplicate and quantified by authentic standards. Results were expressed in µg tocopherol per g of oil.

Secondary Volatiles Oxidation Products (SVOP)—Dynamic Headspace GC-MS

Approximately, 1 g of powder and 30 mg of internal standard (4-methyl-1-pentanol and 30 µg/g water) were weighed in a purge bottle and mixed with 10 mL of distilled water. Then, the bottle content was heated for 30 min in a water bath at 45 °C while purging with nitrogen (flow rate 150 mL/min). The volatile compounds released were retained in Tenax GR tubes (Perkin Elmer 1/4" stainless steel tubes packed with 225 ± 3 mg Tenax GR 60–80 mesh). Then, these compounds were desorbed again by means of helium and heat (200 °C) in an Automatic Thermal Desorber (ATD-400; Perkin Elmer, Norwalk, CN), cryofocused on a cold trap (−30 °C), released again (220 °C) and led to a gas chromatograph (HP 5890IIA; Hewlett Packard, Palo Alto, CA, USA and Column: DB-1701, 30 m × 0.25 mm × 1.0 µm; J&W Scientific, Folsom, CA, USA). The individual compounds were analyzed by mass-spectrometry (HP 5972 mass-selective detector; Agilent Technologies, Santa Clara, California, USA; electron ionization mode: 70 eV and mass to charge ratio scan between 30 and 250). The released volatile compounds were then identified by MS-library searches (Wiley 138K, John Wiley and Sons, Hoboken, New Jersey, USA and Hewlett-Packard, San Jose, California, USA) and quantified through calibration curves using external standards (butanal, pentanal, 1-penten-3-ol, hexanal, (*E*)-2-hexenal, heptanal, octanal, (*E*,*E*)-2,4-heptadienal, (*Z*)-4-heptenal and nonanal) dissolved in 96% ethanol. The standard solutions were diluted to concentrations of approximately 2.5, 5, 10, 50, 100 and 500 µg/mL, and 1 µL of each was directly injected on the Tenax tubes. Measurements were made in triplicates for each sample. Results were expressed as ng/g oil.

2.6. *Production of Fortified Mayonnaise*

Light mayonnaise (40 wt% of total oil) fortified with ω-3 PUFAs (2.5 wt% fish oil) was produced following three different approaches: (i) incorporating neat fish oil (M-NFO), (ii) incorporating a fish oil-in-water emulsion stabilized with WPH (M-EM) and (iii) incorporating spray-dried microcapsules loaded with fish oil and obtained with WPH as a film-forming material and glucose syrup as the encapsulating agent (M-GS). In all cases, 300 g of mayonnaise containing: 2.5 wt% of fish oil, 37.5 wt% of SFO, 4 wt% of egg yolk, 1 wt% of vinegar, 0.4 wt% of lemon juice and 0.3 wt% of salt were prepared as described by Miguel et al. [24], with some modifications. In the case of M-NFO and M-EM samples, 1 wt% of sugar was also added. For the mayonnaise preparation, first, distilled water, salt, sugar and 6 mL of sodium azide solution (0.0125 g/mL) were mixed in a blender (Taurus robot, 300 inox.) for 15 s. For the M-EM sample, also 150 g of emulsified fish oil were added in the first step. Then, the egg yolk was incorporated and mixed for 15 s. Stabilizer Grinsted FF was manually dissolved in 10 g of sunflower oil (and fish oil in the case of the M-NFO sample) and added to the blender (15 s mixing). Then, the rest of the oil was added in three steps (except ca. 10 wt%) and mixed for 30 s each. Vinegar and lemon juice were dispersed manually to the remaining oil (ca. 10 wt%) and added as the last step. Mixing in this case lasted 30 s. In the case of the M-GS sample, 58.5 g of microcapsules were added to the blender at this last step and mixed for 45 s to complete the dispersion.

2.7. Characterization of Fortified Mayonnaise

2.7.1. Physical Stability: Droplet Size Distribution and Viscosity

The droplet size distribution and viscosity of the fortified mayonnaises were evaluated after production and during 28 days of storage at 25 °C.

The droplet size distribution was measured as described in Section 2.4. For this purpose, 1 g of mayonnaise was dissolved in sodium dodecyl sulfate (SDS) buffer (10-mM NaH_2PO_4, pH 7) to a ratio of 1:5 (*w*/*w*). Then, the solution was sonicated for 15 min to avoid droplets agglomeration. Measurements were made in triplicate, and the results are given in surface area mean diameter (D[3,2]) and volume weighted mean diameter (D[4,3]).

The viscosity of the fortified mayonnaise samples was measured using a rotatory Kinexus Malvern rheometer (Malvern Panalytical Ltd., Worcestershire, UK) equipped with a plate-plate geometry. An increasing gradient of stress was applied from 0.1–200 Pa at 25 °C. Measurements were made in triplicate.

2.7.2. Oxidative Stability

To monitor the oxidative stability of low-fat fortified mayonnaise, 40 g of each sample were stored in polyethylene containers (60 mL) at 25 °C during 28 days. Samples were taken at days 0 (week 0), 7 (week 1), 14 (week 2), 21 (week 3) and 28 (week 4) and kept under an inert atmosphere at −80 °C until analysis.

Peroxide Value (PV)

Oil extraction from the low-fat mayonnaise for the PV determination was made as follows: ca. 0.5 g of mayonnaise were weighed and mixed with 5 mL of distilled water. Then, 20 mL of hexane/2-propanol (1:1, *v*/*v*) solvent was added and mixed for 5 min. The resulting mixture was centrifuged at 670× *g* for 2 min. PV was measured using the thiocyanate assay, as described in Drusch et al. [25], with some modifications. Extracted oil was diluted with 2-propanol prior to the addition of iron-II-chloride and ammonium thiocyanate solutions and then was incubated for 5 min at 25 °C. Oil extraction was made in duplicate for each sample. The PV measurements were made in triplicate for each oil extract. Results were expressed in meq O_2 per kg of oil.

P-Anisidine Value (AV)

For the p-anisidine value determination, 2.5 g of mayonnaise were weighed to carry out the oil extraction, as described in Section 2.7.2. Oil extraction was made in duplicate for each sample. The AV measurements were made in triplicate according to the ISO 6885:2006 method [26] for each oil extract.

2.8. Statistical Analysis

For data analysis, the software Statgraphics Centurion XV (Statistical Graphics Corp., Rockville, MD, USA) was used. First, one-way ANOVA was performed to identify if there were differences between the samples. Then, a multiple sample comparison using Tukey's test allowed to identify means which were significantly different from each other. Differences between means were considered significant at $p \leq 0.05$.

3. Results

3.1. Oil Droplet Size Distribution (ODSD) of Emulsions

The parent emulsions and reconstituted emulsions presented values of the 90th percentile below 2 μm (Table 1), indicating that the WPH efficiently stabilized the oil droplets by maintaining the structural integrity of the oil/water interface prior, during and after the microencapsulation process [16]. Despite the lack of surface-active properties of the encapsulating agents used, the results show significant differences ($p < 0.05$) for the ODSD of the parent emulsions prior to drying, which are mainly attributed to minor differences in pressure adjustments in the homogenizer. In this line, Hogan et al. [27] found that the volume average diameter (D[4,3]) of soy oil emulsions stabilized with sodium caseinate did not differ irrespective of the dextrose equivalence (DE) of the non-surface-active carbohydrate used as a wall constituent.

Table 1. Oil droplet size distribution (ODSD) of fresh and reconstituted emulsions during storage time at 4 and 25 °C.

		d_{90}, µm	
-		**GS**	**MD21**
Parent emulsions		0.587 ± 0.001 $^{+,*}$	0.555 ± 0.001 $^{+,*}$
Reconstituted emulsion after spray-drying			
Week 0		0.613 ± 0.006 $^{+,a,u,*}$	0.663 ± 0.001 $^{+,a,u,*}$
4 °C	Week 2	0.624 ± 0.006 ^{a,j,*}	0.699 ± 0.005 ^{b,j,*}
	Week 4	0.653 ± 0.002 ^{b,j,*}	0.730 ± 0.001 ^{c,j,*}
	Week 6	0.654 ± 0.004 ^{b,j,*}	0.825 ± 0.001 ^{d,j,*}
25 °C	Week 2	0.599 ± 0.001 ^{u,k,*}	0.703 ± 0.011 ^{v,j,*}
	Week 4	0.669 ± 0.005 ^{v,j,*}	0.717 ± 0.003 ^{v,k,*}
	Week 6	0.659 ± 0.010 ^{v,j,*}	0.882 ± 0.002 ^{w,k,*}

GS: glucose syrup and MD21: maltodextrin with DE 21. Means within the same column followed by a plus sign, $^{+}$, indicates statistical differences ($p \leq 0.05$) between the parent and reconstituted (week 0) emulsions. Means within the same column followed by different letters, $^{a\text{–}c}$, indicate statistical differences ($p \leq 0.05$) between sampling points for the same encapsulating agent at 4 °C. Means within the same column followed by different letters, $^{u\text{–}w}$, indicate statistical differences ($p \leq 0.05$) between sampling points for the same encapsulating agent at 25 °C. Means within the same column followed by different letters, ^{j,k}, indicate statistical differences ($p \leq 0.05$) between samples stored at different temperatures for the same encapsulating agent at the same sampling point. Means within the same row followed by an asterisk, *, indicates statistical differences ($p \leq 0.05$) between encapsulating agents.

After drying, a small but significant ($p < 0.05$) increase in the ODSD occurred in both samples at week 0, suggesting coalescence of the non-encapsulated oil after reconstitution [28]. In the case of GS-based reconstituted emulsions, the ODSD during six weeks of storage was in the range 0.61–0.66 µm for both storage temperatures. On the other hand, for MD21-based emulsions, the ODSD increased progressively during the storage time from 0.66 to 0.82 µm at 4 °C and 0.66 to 0.88 µm at 25 °C. These results suggest oil leakage in the MD21 microcapsules, which increased the surface fat content and led to more oil available to coalesce after reconstitution. This finding may be related to the poorer retention properties of MD21 as a wall material when compared to GS, attributed to the lower molecular weight of the oligosaccharides present in the glucose syrup. Moreover, the results show that the storage temperature had little effect on the retention properties of the encapsulating agents used, despite the statistical analysis results showing significant differences for some sampling points for both storage temperatures assayed.

3.2. *Physicochemical Characterization of Microencapsulates*

3.2.1. Moisture Content, Water Activity and Encapsulation Efficiency (EE)

The moisture content (MC) was of 4.7% ± 0.4% and 4.2% ± 0.1% for the GS and MD21-based microencapsulates, respectively. These low MC values, together with the low water activity (a_w) of the microencapsulates (GS, 0.184 ± 0.004 and MD21, 0.187 ± 0.001), are desired in order to confer long-term microbiological stability for the microcapsules. Moreover, as no significant differences could be observed between the samples ($p > 0.05$), it can be assumed that the MC and a_w are independent of the encapsulating agent used (while maintaining constant the spray-drying processing variables).

High EE values were obtained for both types of microcapsules, although a significantly higher EE value was found for the GS-based microencapsulates when compared to the MD21-based microcapsules (98.07 ± 0.04% vs. 97.66 ± 0.06%). Interestingly, the slightly higher EE value of the GS-based microencapsulates cannot be attributed to a smaller droplet size of the parent emulsion (Table 1). Thus, differences in the EE may be consequence of a positive correlation between the EE of encapsulates and DE of the carbohydrates used as wall materials [27,29,30]. This fact is explained on the basis that increasing the DE of the carbohydrate leads to smaller oligosaccharides, which are thought to form a more uniform and denser packaging of the core material, which favors oil encapsulation.

3.2.2. Morphology and Size

Figure 1 shows the morphology and particle size distribution of the microcapsules. Both types of microcapsules showed a spherical shape with smooth surfaces, even though a few microcapsules with wrinkled surfaces could also be observed. Furthermore, no particle agglomerations were detected, which can be attributed to the low surface fat content of the microencapsulates produced [27]. Microcapsules prepared with MD21 presented a wider particle size distribution when compared to the microcapsules produced with GS. The percentage of microcapsules with diameters below 20 µm was 92.9% and 70.5% for the GS and MD21-based microencapsulates, respectively (Figure 1), while the particle mean diameters were 12.5 ± 5.5 µm and 16.8 ± 10.2 µm, respectively. Since the rotational speed of the atomizer was kept constant, the larger particle size of the MD21-based microencapsulates was most likely due to the higher viscosity of the MD21-feed emulsion, which led to larger emulsion droplets after atomization and, hence, larger dried microcapsules when compared to the GS-feed emulsion [27,31].

3.2.3. Oxidative Stability of the Microencapsulates

Peroxide Value (PV) and Tocopherol Content (TC)

Oxidative stability of the microencapsulates was first evaluated by determining the formation of the primary oxidation products during storage (Figure 2). Right after production, PV of 3.24 ± 1.38 and 3.73 ± 0.66 meq O_2/kg oil were obtained for the GS and MD21-based microcapsules, respectively. No significant differences ($p > 0.05$) in the PV were observed between the two types of microcapsules at day 0, although a significant increase on the hydroperoxide content compared to the fresh oil (PV = 0.36 ± 0.03 meq O_2/kg oil) could be noted. This is attributed to the encapsulation process itself, which involves mechanical stress, shear forces and heat, leading to the initial oil oxidation [8,9]. As expected, lipid oxidation was accelerated for the two types of microcapsules when increasing the storage temperature from 4 to 25 °C. For lipid autoxidation, an increase of 10 °C temperature results in a two-fold increase in the reaction rate [32]. However, the effect was more pronounced in the MD21-based microcapsules (PV = 9.01 ± 0.13 meq O_2/kg oil after six weeks storage at 25 °C). This positive correlation between the lipid oxidation rate and storage temperature on microencapsulated fish oil has also been reported by other authors [33,34] and could be attributed to the loss of natural antioxidants, as well as an increase in diffusivity of the prooxidant agents (e.g., oxygen, radicals and trace of metals) at higher temperatures.

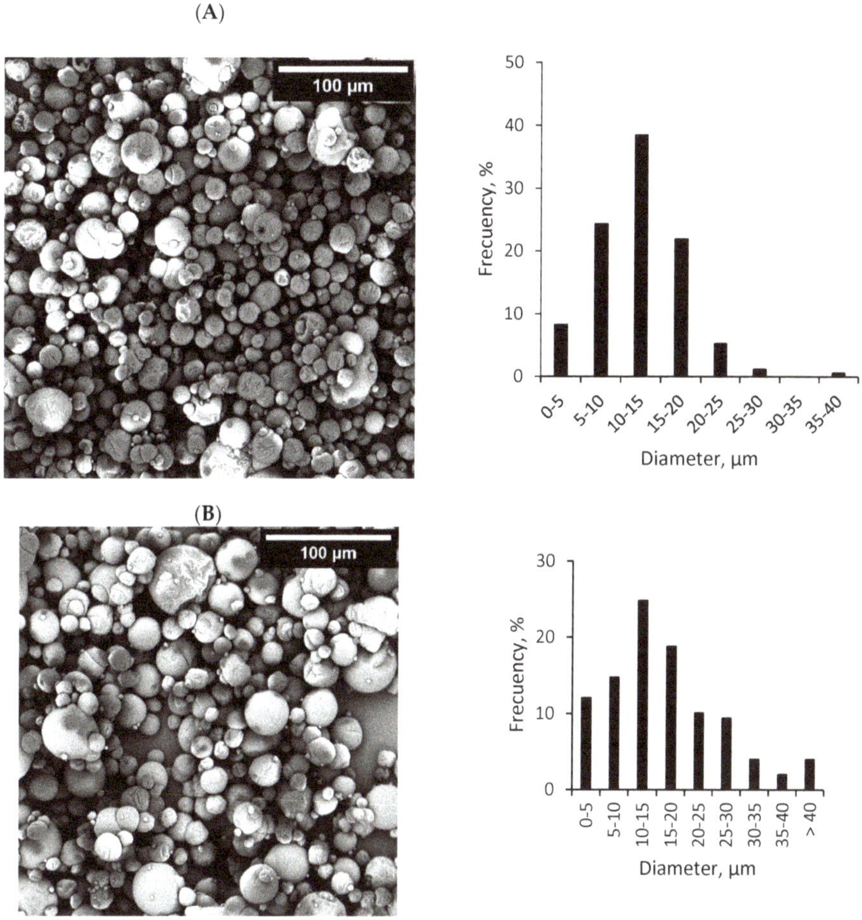

Figure 1. SEM micrographs and diameter distribution of fish oil-loaded microcapsules containing whey protein hydrolysate (WPH) as a film-forming material and glucose syrup (**A**) or maltodextrin DE21 (**B**) as the encapsulating agents.

Nonetheless, although the temperature influenced lipid oxidation, higher PV values were found in MD21-based microcapsules in all cases. These results indicate that the influence of the wall material (e.g., which determines the permeability to prooxidant species and retention properties) predominated over the storage temperature on microencapsulated fish oil oxidative stability. The lower PV of the GS-based microencapsulates has been attributed to the DE differences between carbohydrates and, thus, the molecular weight. Such differences affect the matrix free volume elements and the oxygen diffusivity through the wall. The latter increases with the increasing molecular weight of the carbohydrate [35,36]. Moreover, the ODSD results showed oil leakage during the storage time for MD21-based microencapsulates (Section 3.1), which resulted in lower EE for MD21-based microcapsules and increased the content of easily oxidizable oil on the surface of these microcapsules.

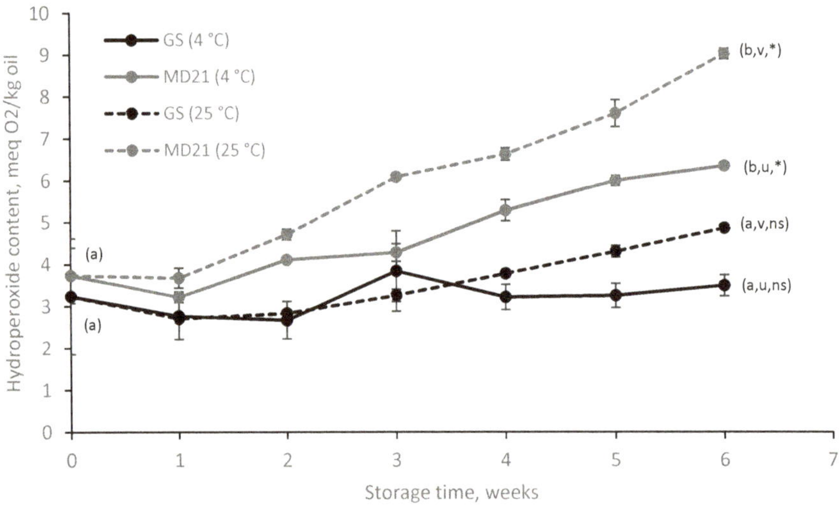

Figure 2. Peroxide value (PV) of spray-dried microcapsules loaded with fish oil during storage at: 4 °C (solid line, —•—) and 25 °C (broken line, - -•- -) encapsulated with glucose syrup (black) or maltodextrin (grey). Means within the same sampling point followed by a letter, a-b, indicates statistical differences ($p \leq 0.05$) between encapsulating agents for the same storage temperature. Means within the same sampling point followed by a letter, u-v, indicates statistical differences ($p \leq 0.05$) between storage temperatures for the same encapsulating agent. Means within the same sample followed by an asterisk (*) indicates statistical differences ($p \leq 0.05$) between week 0 and week 6. Means within the same sample followed by "ns" indicates no statistical differences ($p > 0.05$) between week 0 and week 6.

Moreover, it is noteworthy that the PV results of this study were lower than others reported in literature that also produced fish oil microencapsulates within carbohydrate matrices. For instance, Drusch and Berg [37] produced fish oil-loaded microcapsules (30 wt% or 50 wt%) by spray-drying into a matrix of nOSA-starch and GS at two drying temperature settings (160/60 °C or 210/90 °C) with PV of ca. 160–340 meq O_2/kg oil after six weeks of storage at 20 °C. Likewise, Morales-Medina et al. [9] microencapsulated fish oil using fish protein hydrolysates as emulsifiers and GS as the encapsulating material. The PV of the microencapsulates reported by the authors after six weeks of storage at 20 °C were also higher than those values obtained in the current study (ca. 120–160 meq O_2/kg oil). It should be noted that both the GS used and the inlet/outlet drying temperatures were similar to the present work.

In addition to the PV, alpha-, beta-, delta- and gamma-tocopherols were quantified in the microcapsules during storage (see Figure S1 in the Supplementary Material). A significant decrease ($p < 0.05$) was observed in the content of alpha-, delta- and gamma-tocopherols for the microcapsules when compared to the fresh oil. This indicates the formation of radicals during emulsification and drying processes and that tocopherols were consumed as a consequence of their radical scavenging activity. It is worth mentioning that the high initial content of tocopherols in the fresh fish oil is a consequence of the addition of these compounds to the oil by the producer. The tocopherol content of the microencapsulates slightly changed over the storage time, regardless of the encapsulating agent used and the storage temperature. The latter leads us to conclude that the oxidative stability of the microcapsules during storage was not influenced by their tocopherol content but was mainly determined by the protection provided by the encapsulating agents, as well as the stabilization provided with the WPH.

Secondary Volatile Oxidation Products (SVOP)

Figure 3 shows the content of the SVOP (1-penten-3-ol, hexanal, (E)-2-hexenal, (Z)-4-heptenal and nonanal) in the microcapsules during storage. Despite the fact that hexanal and nonanal are compounds mainly derived from the oxidation of omega-6 and omega-9 fatty acids, these compounds have also been identified in oxidized fish oil. The rest of the identified SVOP (1-penten-3-ol, (E)-2-hexenal and (Z)-4-heptenal) are typical secondary oxidation products derived from omega-3 PUFAs, and all are related to undesirable odors and flavors [38]. For instance, the odor of 1-penten-3-ol, hexanal and (E)-2-hexenal have been described as earthy and grassy, while (Z)-4-heptenal and nonanal have been described as creamy and tallowy, respectively [39,40]. The odor threshold values for these compounds in oxidized oils range from 0.04 to 3.0 ppm [38].

As occurred with the PV, the higher storage temperature favored lipid oxidation for both types of microencapsulates, since the highest SVOP content was found in the samples stored at 25 °C compared to those stored at 4 °C. However, different trends in the SVOP curves can be observed during the storage time (Figure 3). In the case of 1-penten-3-ol (Figure 3A) and (Z)-4-heptenal (Figure 3D), the influence of the storage temperature predominated over the wall material regarding lipid oxidation, since the lower contents of these compounds were found in the encapsulates stored at 4 °C, regardless of the encapsulating agent. Nonetheless, it is noteworthy that, in both cases, the GS-based microcapsules were the less-oxidized after six weeks of storage (48.06 ± 9.57 ng of 1-penten-3-ol/g oil and 14.85 ± 2.14 ng of (Z)-4-heptenal/g oil). On the contrary, the trend of hexanal (Figure 3B), (*E*)-2-hexenal (Figure 3C) and nonanal (Figure 3E) curves are in line with the PV results (Section 3.2.3, with MD21-based microcapsules presenting higher contents of the these volatiles independently of the storage temperatures when compared to GS-based microcapsules. This indicates, for the latter SVOP, a clear influence of the wall material over the storage temperature on microencapsulated fish oil oxidative stability.

Most of the studies published regarding the oxidative stability of microencapsulated fish oil by spray-drying have only analyzed the course of propanal during storage, since it is an important SVOP resulting from the oxidation of both EPA and DHA [8,35,37,41]. However, our results show that different trends can be observed when studying more than one SVOP. Likewise, García-Moreno et al. [19] studied the course of 1-penten-3-ol, (E)-2-pentenal, heptanal and nonanal during 21 days of storage at 20 °C for electrosprayed microcapsules loaded with fish oil (20 wt%) into a matrix of pullulan blended with dextran or GS. For 1-penten-3-ol, the authors reported values ranging from 500–2000 ng/g oil, while, for nonanal, the values ranged from 2000–5000 ng/g oil. In any case, the values reported by these authors were much higher than those of the current work, which are well below the compounds' odor threshold limits in oil, indicating that the microcapsules may be stable in regard to their organoleptic properties.

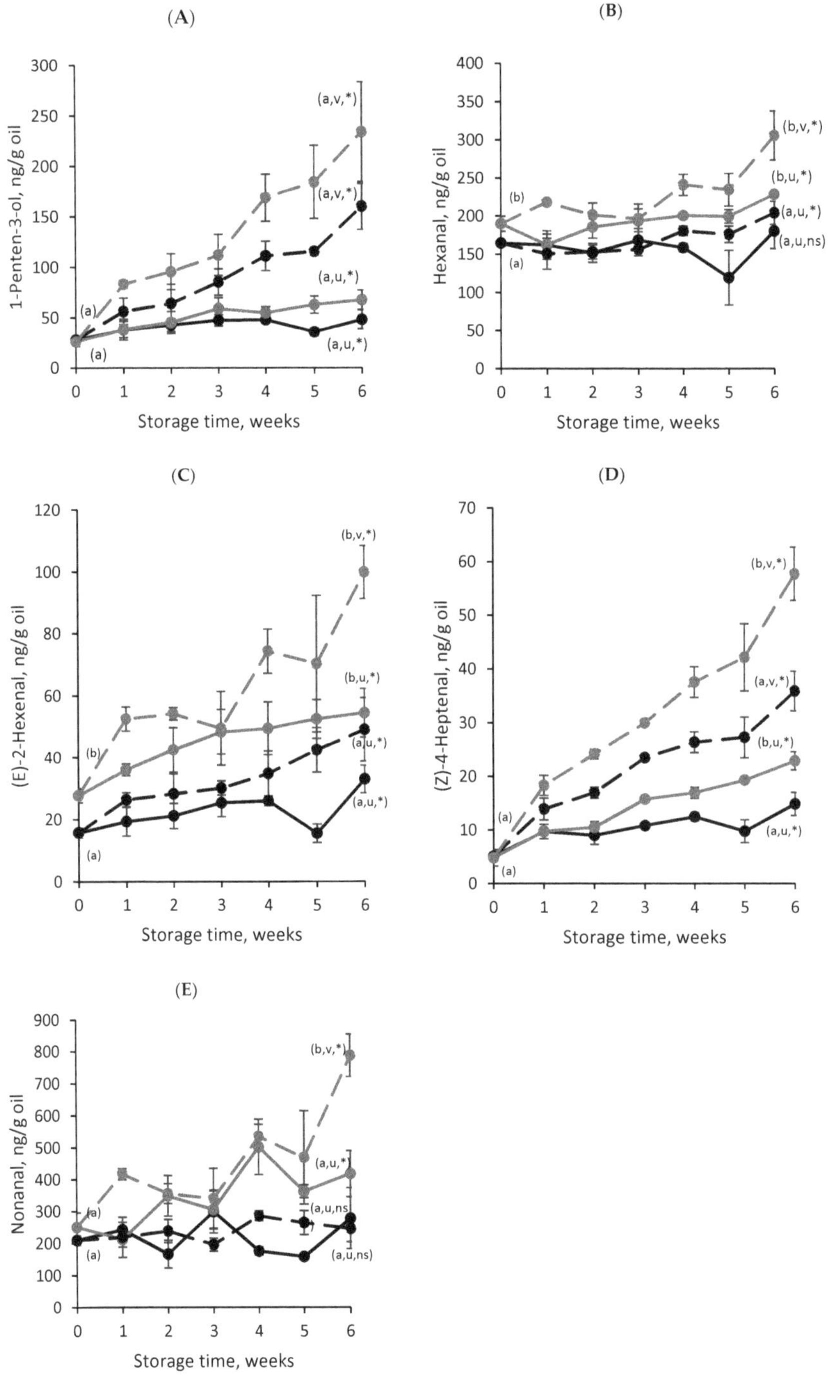

Figure 3. Secondary volatile oxidation products (1-penten-3-ol (**A**), hexanal (**B**), (**E**)-2-hexenal (**C**), (Z)-4-heptenal (**D**) and nonanal (**E**)) of spray-dried microcapsules loaded with fish oil during storage at:

4 °C (solid line, —•—) and 25 °C (broken line, - -•- -) encapsulated with glucose syrup (black) and maltodextrin (grey). Means within the same sampling point followed by a letter, a-b, indicates statistical differences ($p \leq 0.05$) between encapsulating agents for the same storage temperature. Means within the same sampling point followed by a letter, u-v, indicates statistical differences ($p \leq 0.05$) between storage temperatures for the same encapsulating agent. Means within the same sample followed by an asterisk (*) indicates statistical differences ($p \leq 0.05$) between week 0 and week 6. Means within the same sample followed by "ns" indicates no statistical differences ($p > 0.05$) between week 0 and week 6.

Taken altogether, the PV and SVOP content results show that the most oxidatively stable microcapsules were those produced with GS as a wall material. This is directly related to the lower molecular weight of GS as a result of its higher DE when compared to MD21, which led to microencapsulates: (i) with higher EE and retention of the core material and (ii) less porous and, thus, less permeable to prooxidant species (e.g., oxygen). However, the higher oxidative stability of the GS-based microencapsulates reported in the current study cannot only be attributed to the protective effect of the wall material. Tamm et al. [16] produced microcapsules (ca. 31 wt% oil load) spray-drying emulsions stabilized with β-LG hydrolysates (2.2 wt%) and GS as wall material. The authors reported a higher oxidative stability (based on the PV) of the microencapsulates containing hydrolysates when compared to those containing unmodified protein, which clearly revealed the additional protective effect of using whey protein hydrolysates for microencapsulating fish oil. These results are in line with those of our study, where the content of both the hydroperoxides (PV) and SVOP of GS-based microcapsules containing WPH as a film-forming material remained low during six weeks of storage. The protective effect of the WPH used in this study is related to its high emulsifying and antioxidant activities (both radical scavenging, EC_{50} = 4.45 ± 0.00 mg/mL, and metal chelating, EC_{50} = 0.95 ± 0.01 mg/mL), as reported in our previous work (unpublished results). The antioxidant activity of the WPH is mainly attributed to its high content in Tyr, Met and, to a lesser extent, His. Tyr has been related to possess radical scavenging activity due to the capacity of the phenolic group to serve as the hydrogen donor [12,42], Met reduces lipid hydroperoxides to the non-reactive species by oxidation to Met sulfoxide [12,43] and His possesses both radical scavenging and chelating activity [44]. Moreover, the small peptides produced as a result of protein hydrolysis (ca. 50% of the peptides between 0.5–3 kDa, (unpublished results) could have also improved the encapsulating matrix protective effect by acting as copolymers or fillers of the wall, thus limiting prooxidant diffusivity to a higher extent. This fact combined with an already low permeable wall matrix (e.g., GS) and low storage temperatures resulted in long-term oxidatively stable microcapsules.

Hence, GS-based microcapsules loaded with fish oil and containing WPH as a film-forming material were further evaluated as an omega-3 delivery system to produce fortified low-fat mayonnaise.

3.3. *Physical and Oxidative Stabilities of Fortified Low-Fat Mayonnaise*

The feasibility of using fish oil-loaded microcapsules produced by spray-drying (using GS as the encapsulating agent and WPH as the film-protein material) to enrich low-fat mayonnaise was evaluated. For this purpose, the physical and oxidative stabilities of low-fat mayonnaise samples fortified with: (i) neat fish oil (M-NFO), (ii) emulsified fish oil (M-EM) and (iii) microencapsulated fish oil (M-GS) were investigated.

3.3.1. Physical Stability: Droplet Size Distribution and Viscosity

Droplet size distributions of the mayonnaise samples indicate their physical stability but also the specific surface area, which influences lipid oxidation in the system. After production (day 0), the mayonnaise enriched with neat fish oil (M-NFO) showed a bimodal droplet size distribution with a main peak centered in ca. 1 μm and a second one, less representative, centered in ca. 4.7 μm (Figure 4A). Likewise, the mayonnaise sample containing microencapsulated fish oil (M-GS) showed a bimodal droplet size distribution with the peaks centered in 0.13 and 1.8 μm, respectively (Figure 4A).

When comparing this curve to that of the emulsion used for the production of the microencapsulates, the droplet size of the first peak of the mayonnaise is in agreement with the peak of the monomodal curve of the emulsion fed to the spray-drier (ca. 0.13 μm, data not shown). Therefore, the first peak of the M-GS sample represents the droplet size of the fish oil, while the second peak is a consequence of the sunflower oil droplets dispersed during the mayonnaise production. The mayonnaise fortified with emulsified fish oil (M-EM) showed a trimodal droplet size distribution with the first peak overlapping with the M-GS sample (ca. 0.13 μm) and the other two peaks centered as those of the M-NFO sample (ca. 1 and 4.7 μm, respectively) (Figure 4A). The latter suggests that the fish oil emulsion maintained its physical integrity during the mayonnaise production process, despite the shear forces produced by the blender, and did not modify the sunflower oil droplet size distribution.

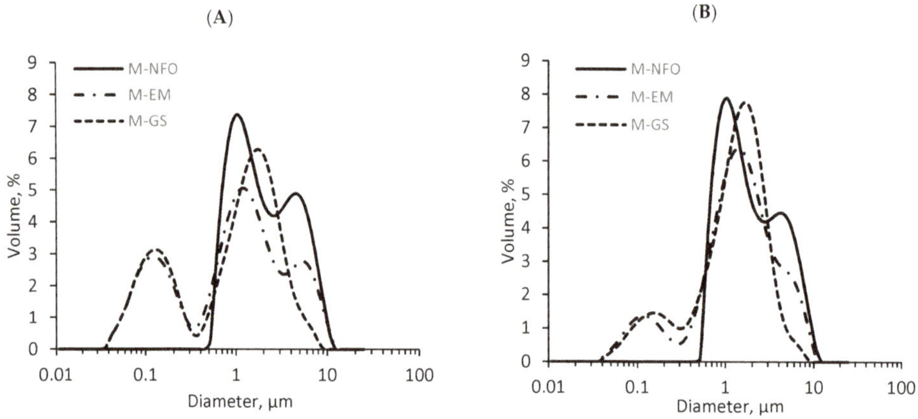

Figure 4. Droplet size distribution of low-fat mayonnaise enriched with: (i) neat fish oil (M-NFO), (ii) emulsified fish oil (M-EM) and (iii) microencapsulated fish oil (M-GS) at (**A**) day 0 and (**B**) day 28.

Table 2 shows that the mayonnaise sample containing neat fish oil (M-NFO) presented D[4,3] and D[3,2] values significantly higher ($p < 0.05$) than the rest of the samples at day 0. This is because, for the M-EM and M-GS samples, the fish oil was emulsified using a high-pressure homogenizer prior to the addition to the mayonnaise, which resulted in a lower size of the fish oil droplets when compared to the complete homogenization of the sunflower and fish oils while using the kitchen blender. It should be noted that, for the M-GS sample, the microcapsules were dispersed for 45 s into an already produced mayonnaise, which could have led to a further decrease of the size of the bulk droplets (sunflower oil droplets), explaining the lower D[4,3] values observed for this sample. Contrary to other studies, which found that the fortified mayonnaise with microencapsulated fish oil was the one with a higher D[4,3] value [24,45], our results may indicate that the capsules were intact and well-dispersed in the mayonnaise matrix at day 0. It should be noted that the differences in the D[4,3] values between our results and those reported in the previous studies mentioned [24,45] are mainly attributed to the different equipment used for mayonnaise production.

Table 2. Droplet size and apparent viscosity ($\gamma = 10\ s^{-1}$) of fortified low-fat mayonnaise enriched with neat fish oil (M-NFO), emulsified fish oil (M-EM) or microencapsulated fish oil (M-GS).

	Droplet Size				**Apparent Viscosity**	
	Day 0		**Day 28**		**($\gamma = 10\ s^{-1}$), Pa·s**	
Sample	**D[3,2], μm**	**D[4,3], μm**	**D[3,2], μm**	**D[4,3], μm**	**Day 0**	**Day 28**
M-NFO	1.555 ± 0.020 ^{a}	2.724 ± 0.035 ^{a}	1.487 ± 0.006 ^{a,*}	2.553 ± 0.011 ^{a,*}	2.6 ± 0.3^{a}	2.0 ± 0.1 ^{a,*}
M-EM	0.310 ± 0.008 ^{b}	1.865 ± 0.052 ^{b}	0.516 ± 0.038 ^{b,*}	1.985 ± 0.045 ^{b,ns}	4.8 ± 0.1^{b}	5.1 ± 0.5 ^{b,ns}
M-GS	0.302 ± 0.001 ^{b}	1.436 ± 0.039 ^{c}	0.519 ± 0.027 ^{b,*}	1.538 ± 0.068 ^{c,ns}	9.7 ± 0.3^{c}	9.4 ± 0.2 ^{c,ns}

Means within the same column followed by a letter, $^{a-c}$, indicates statistical differences ($p \leq 0.05$) between samples. Means within the same sample followed by an asterisk, *, indicates statistical differences ($p \leq 0.05$) between day 0 and day 28. Means within the same sample followed by "^{ns}" indicates no statistical differences ($p > 0.05$) between day 0 and day 28.

After 28 days of storage, fortified mayonnaise samples retained their physical integrity, since no phase separation or creaming were observed. However, changes in the droplet size distribution occurred (Figure 4B and Table 2). Contrary to the M-EM and M-GS samples, the D[4,3] and D[3,2] values of M-NFO decreased after 28 days of storage, suggesting that right after mayonnaise production, oil floccules were present, which may have disintegrated during storage [24]. For the M-EM and M-GS samples, a significant increase in the D[3,2] value suggested the coalescence or flocculation of oil droplets during storage caused by: (i) partial physical destabilization of the fish oil-in-water emulsion (in the case of the M-EM sample) or (ii) partial disintegration of the encapsulating wall, resulting in oil release (in the case of the M-GS sample) when these delivery systems were added to the food matrix.

All fortified low-fat mayonnaise samples presented pseudoplastic behaviors (see Figure S2 in the Supplementary Material). Table 2 shows the apparent viscosity values at a shear rate of 10 s^{-1} for the mayonnaise samples. It was observed that the M-GS presented a significantly higher apparent viscosity, followed by the M-EM and M-NFO mayonnaises. These results are in line with those previously reported by Miguel et al. [24], who attributed the higher viscosity of the mayonnaise sample enriched with zein-based fish oil microencapsulates to the intact microcapsules' thickening effect in the aqueous phase. Moreover, it has been stated that the larger the oil droplet size, the lower the apparent viscosity, since a reduced surface-to-volume ratio of the disperse phase leads to less friction between the droplets [46]. The latter correlates well with these results, since as the droplet size of the fortified mayonnaise samples increased (e.g., higher for the M-NFO), their apparent viscosity decreased (e.g., lower for the M-NFO) (Table 2). After 28 days of storage, the apparent viscosity of the M-EM and M-GS samples did not change significantly ($p > 0.05$), contrary to the M-NFO sample. This may be explained by the presence of WPH in the M-EM and M-GS mayonnaises, the latter also containing GS. This fact resulted in an improved physical stability of the M-EM and M-GS samples, which was attributed to an increased viscosity [47].

Overall, the GS-based enrichment system led to a fortified mayonnaise with a high physical stability caused by an increased viscosity.

3.3.2. Oxidative Stability of the Fortified Mayonnaise

Peroxide Value (PV)

Figure 5 shows the evolution of the PV for the enriched low-fat mayonnaise samples during storage. It was observed that the PV at day 0 was similar among the three mayonnaise samples (4.0 ± 0.2 to 4.6 ± 0.1 meq O_2/kg oil), despite the different degrees of initial oxidation of the enrichment systems: (i) NFO (PV = 1.3 ± 0.4 meq O_2/kg oil; AV = 6.6 ± 0.7), (ii) fish oil-in-water emulsion stabilized with WPH (PV = 4.8 ± 1.2 meq O_2/kg oil; AV = 6.2 ± 1.1) and (iii) fish oil-loaded glucose syrup microcapsules containing WPH as a film-forming material (PV = 7.0 ± 1.0 meq O_2/kg oil; AV = 11.9 ± 1.5). This can be explained, because the majority of the oil present in the mayonnaise was

SFO (PV = 3.5 ± 0.2 meq O_2/kg oil; AV = 8.1 ± 1.1), which is more oxidatively stable than fish oil and may have masked the contribution of fish oil oxidation in the mayonnaise samples after production.

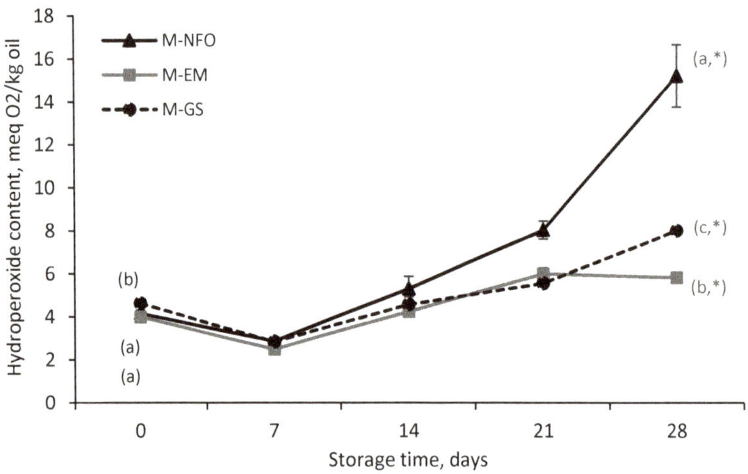

Figure 5. Peroxide value (PV) of low-fat fortified mayonnaise enriched with neat fish oil (M-NFO), emulsified fish oil (M-EM) or microencapsulated fish oil (M-GS) over storage time at 25 °C. Means within the same sampling point followed by a letter, a-c, indicates statistical differences ($p \leq 0.05$). Means within the same sample followed by an asterisk (*) indicates statistical differences ($p \leq 0.05$) between day 0 and day 28.

Over storage time, the PV course showed a lag phase of 1 week for the three enriched mayonnaise samples (Figure 5). However, from week 2 onwards, the PV of the three samples assayed increased progressively to final values significantly different ($p < 0.05$) from those of the beginning of the storage time. Interestingly, at day 28, the mayonnaise fortified with NFO presented a significantly higher PV (15.22 ± 1.45 meq O_2/kg oil) compared to M-GS and M-EM. It should be also noted that, although the final PV of M-GS was significantly higher than the PV of M-EM, a slight decrease in the PV of M-EM was observed after day 21. This indicates that the rate of hydroperoxide decomposition was higher compared to the rate of hydroperoxide formation after day 21 for M-EM.

It is also noteworthy that our results are not aligned with those reported by Hermund et al. [45]. These authors found a lower oxidative stability for low-fat mayonnaise enriched with GS-based electrosprayed microcapsules (PV = 8–12 meq O_2/kg oil) when compared to low-fat mayonnaise enriched with NFO (PV = 2 meq O_2/kg oil) after 28 days storage at 25 °C. This finding was attributed to the partial disintegration of the hydrophilic biopolymer-based wall matrix of the electrosprayed microcapsules during mayonnaise production, which led to the release of already oxidized fish oil. On the other hand, Miguel et al. [24] produced fortified mayonnaise enriched with zein-based electrosprayed microcapsules with a high oxidative stability during storage (PV = 2 meq O_2/kg oil after 21 days of storage at 25 °C). In the latter work, the authors confirmed that the microencapsulates remained intact after mayonnaise production, which explained the enhanced oxidative stability of the fortified product. Taking both studies' results into account leads us to speculate that, in the current study, the physical integrity of the encapsulating wall was not severely affected either by the fortified mayonnaise production process (high-speed blending to disperse the capsules) nor the food matrix (water-based emulsion). However, the latter requires further investigation.

P-Anisidine Value (AV)

The p-anisidine value (AV) indicates the formation of secondary oxidation products, principally 2-alkenals and 2,4-alkadienals, which arise as a consequence of decomposition of the primary oxidation products (hydroperoxides) [48]. The initial AV of the fortified mayonnaises ranged from 6.2 ± 1.4 to 6.9 ± 0.3, with no significant differences among the samples ($p > 0.05$) (Figure 6).

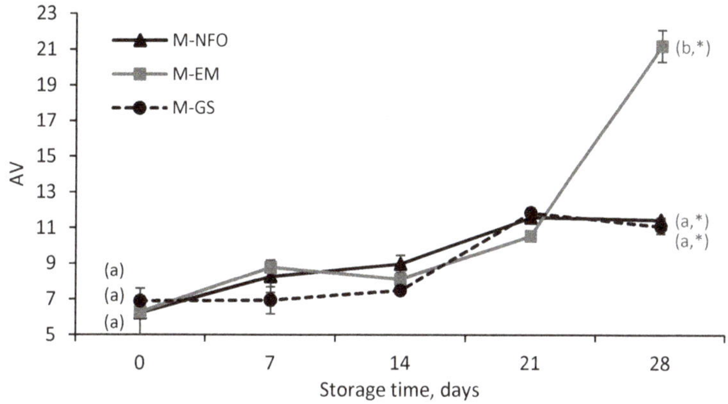

Figure 6. P-anisidine value (AV) of low-fat fortified mayonnaise enriched with neat fish oil (M-NFO), emulsified fish oil (M-EM) or microencapsulated fish oil (M-GS) over storage time at 25 °C. Means within the same sampling point followed by a letter, a-c, indicates statistical differences ($p \leq 0.05$). Means within the same sample followed by an asterisk, *, indicates statistical differences ($p \leq 0.05$) between day 0 and day 28.

During storage, the AV increased progressively for all the samples until day 21, from which a sharp increase in only the M-EM sample occurred. These results are in line with those previously reported for PV since, after day 21, the hydroperoxide content of the M-EM sample decreased (Figure 5), suggesting the appearance of secondary oxidation products. The poorer oxidative stability of the M-EM sample could be result of the emulsification process itself (production of a coarse emulsion and subsequent homogenization). The first step involves intense mechanical stress, which favors air inclusion, while the latter disrupts/rearranges the oil droplets, favoring a better distribution of the prooxidant species within the matrix (e.g., oxygen or metal ions) [8]. Moreover, high-pressure homogenization causes a significant increase in a specific surface area, as denoted by the lower D[4,3] values of M-EM when compared to M-NFO (Table 2), which results in a larger surface of contact between the oil and prooxidants. Likewise, Let at al. [49] reported a lower oxidative stability (based on the content of 2-hexenal, 4-heptenal and 2,4-heptadienal during storage) for a salad dressing enriched with emulsified fish oil when compared to the addition of neat fish oil. The authors also reported that the oxidative stability of neat fish oil-fortified yogurt was higher than that of the yogurt fortified with emulsified fish oil.

After 28 days of storage, no significant differences were observed in the AV when comparing the M-NFO and M-GS samples ($p > 0.05$). However, M-GS showed a lag phase of one week, contrary to M-NFO (Figure 6), which could suggest a protective effect of the microencapsulates against fish oil oxidation during the early stages of storage. In this line, Nielsen and Jacobsen [50] investigated the fish oil-enrichment approach for a fish oil-fortified cod pâté and reported that the addition of microencapsulated fish oil resulted in a better oxidative stability after storage (based on 1-penten-3-oil content) than the samples fortified either with emulsified fish oil or neat fish oil.

Taken altogether, the PV and AV results indicate that the most oxidatively stable mayonnaise sample after one month of storage was M-GS, closely followed by M-NFO, while M-EM showed the lowest oxidative stability. Hence, the enhanced protection against lipid oxidation could be the result of the maintenance of the physical integrity of the encapsulating wall during mayonnaise production, which provides a physical barrier limiting the contact between fish oil droplets and prooxidants. Moreover, the higher apparent viscosity of M-GS as a result of the thickening effect of the microencapsulates may have also limited prooxidant species diffusivity, reducing lipid oxidation [51].

Our findings revealed that the use of WPH as a film-forming material, besides its metal chelating activity, which is of special interest to reduce lipid oxidation in egg yolk-based products [45], was only beneficial for the production of microcapsules but not emulsions to be used as omega-3 delivery systems. Hence, the processing stages required to obtain the fish oil-loaded microcapsules containing WPH as a film-forming material (e.g., homogenization and spray-drying) are justified, since they result in an omega-3 delivery system that better protects the fish oil when it is incorporated into low-fat mayonnaise.

4. Conclusions

Oxidatively stable microcapsules loaded with fish oil (ca. 13 wt%) were produced via spray-drying using either GS or MD21 as wall materials and WPH as a film-forming material. GS-based microcapsules showed a higher oxidative stability when compared to MD21-based microcapsules at the two storage temperatures assayed (4 and 25 °C), which was mainly attributed to the lower oxygen diffusivity in GS-microcapsules. However, the higher oxidative stability for the GS-based microcapsules obtained, when compared to previous studies, also suggested a positive influence of the use of WPH as a film-forming material exhibiting antioxidant activity. Moreover, our results showed that the delivery system used to enrich low-fat mayonnaise played a major role on the oxidative stability of the fortified product. Adding the fish oil in the form of a fish oil-in-water emulsion stabilized with WPH-favored lipid oxidation in the enriched product when compared to the addition of either neat or microencapsulated fish oil. This may be explained by successive emulsification processes that contributed to dispersing oxygen and prooxidant agents within the matrix, as well as to increasing the interfacial area. Fortified mayonnaise with GS-based microcapsules loaded with fish oil and containing WPH as a film-forming material showed the higher oxidative stability after storage. This was mainly explained because the physical integrity of the microencapsulates may have remained intact after mayonnaise production. Thus, our results show the potential of using WPH as a film-forming material for the production of fish oil-loaded microcapsules to be used as omega-3 delivery systems.

Supplementary Materials: The following are available online at http://www.mdpi.com/2304-8158/9/5/545/s1: Figure S1: Tocopherol content of spray-dried capsules loaded with fish oil during storage at: 4 °C (solid line, —•—) and 25 °C (broken line, - -•- -) microencapsulated with glucose syrup (black) or maltodextrin (grey); Figure S2: Viscosity of mayonnaise enriched with: (i) neat fish oil (M-NFO), (ii) emulsified fish oil (M-EM) and (iii) microencapsulated fish oil (M-GS) at (A) day 0 and (B) day 28.

Author Contributions: Conceptualization, P.J.G.-M., F.J.E.-C. and E.M.G.; data curation, N.E.R.-M. and I.G.-S.; funding acquisition, E.M.G.; investigation, N.E.R.-M. and I.G.-S.; resources, C.J.; supervision, P.J.G.-M. and F.J.E.-C.; writing—original draft, N.E.R.-M. and I.G.-S. and writing—review and editing, P.J.G.-M., F.J.E.-C., C.J. and E.M.G. All authors have read and agreed to the published version of the manuscript.

Funding: This research was funded by the project CTQ2017-87076-R from the Spanish Ministry of Science, Innovation and Universities. N. E. Rahmani-Manglano acknowledges a FPI grant from the Spanish Ministry of Science, Innovation and Universities (PRE2018-084861).

Acknowledgments: The authors are very grateful to Lis Berner for her skillful help with the PV and SVOP measurements of the microcapsules.

Conflicts of Interest: The authors declare no conflicts of interest.

References

1. Jacobsen, C. Enrichment of foods with omega-3 fatty acids: A multidisciplinary challenge. *Ann. N. Y. Acad. Sci.* **2010**, *1190*, 141–150. [CrossRef]
2. Calder, P.C. Very long chain omega-3 (n-3) fatty acids and human health. *Eur. J. Lipid Sci. Technol.* **2014**, *116*, 1280–1300. [CrossRef]
3. Jacobsen, C. Fish Oils: Composition and Health Effects. In *Encyclopedia of Food and Health*; Elsevier: Oxford, UK, 2016; pp. 686–692.
4. Arab-Tehrany, E.; Jacquot, M.; Gaiani, C.; Imran, M.; Desobry, S.; Linder, M. Beneficial effects and oxidative stability of omega-3 long-chain polyunsaturated fatty acids. *Trends Food Sci. Technol.* **2012**, *25*, 24–33. [CrossRef]
5. Desai, K.G.H.; Park, H.J. Recent developments in microencapsulation of food ingredients. *Dry. Technol. Int. J.* **2005**, *23*, 1361–1394. [CrossRef]
6. Jacobsen, C.; García-Moreno, P.J.; Mendes, A.C.; Mateiu, R.V.; Chronakis, I.S. Use of Electrohydrodynamic Processing for Encapsulation of Sensitive Bioactive Compounds and Applications in Food. *Annu. Rev. Food Sci. Technol.* **2018**, *9*, 525–549. [CrossRef] [PubMed]
7. Encina, C.; Vergara, C.; Giménez, B.; Oyarzún-Ampuero, F.; Robert, P. Conventional spray-drying and future trends for the microencapsulation of fish oil. *Trends Food Sci. Technol.* **2016**, *56*, 46–60. [CrossRef]
8. Serfert, Y.; Drusch, S.; Schwarz, K. Chemical stabilisation of oils rich in long-chain polyunsaturated fatty acids during homogenisation, microencapsulation and storage. *Food Chem.* **2009**, *113*, 1106–1112. [CrossRef]
9. Morales-Medina, R.; Tamm, F.; Guadix, A.M.; Guadix, E.M.; Drusch, S. Functional and antioxidant properties of hydrolysates of sardine (*S. pilchardus*) and horse mackerel (*T. mediterraneus*) for the microencapsulation of fish oil by spray-drying. *Food Chem.* **2016**, *194*, 1208–1216. [CrossRef]
10. McClements, D.J.; Decker, E. Interfacial Antioxidants: A Review of Natural and Synthetic Emulsifiers and Coemulsifiers That Can Inhibit Lipid Oxidation. *J. Agric. Food Chem.* **2018**, *66*, 20–25. [CrossRef]
11. Shahidi, F.; Zhong, Y. Revisiting the polar paradox theory: A critical overview. *J. Agric. Food Chem.* **2011**, *59*, 3499–3504. [CrossRef]
12. Hernández-Ledesma, B.; Dávalos, A.; Bartolomé, B.; Amigo, L. Preparation of antioxidant enzymatic hydrolysates from α-lactalbumin and β-lactoglobulln. Identification of active peptides by HPLC-MS/MS. *J. Agric. Food Chem.* **2005**, *53*, 588–593. [CrossRef] [PubMed]
13. Eberhardt, A.; López, E.C.; Ceruti, R.J.; Marino, F.; Mammarella, E.J.; Manzo, R.M.; Sihufe, G.A. Influence of the degree of hydrolysis on the bioactive properties of whey protein hydrolysates using Alcalase®. *Int. J. Dairy Technol.* **2019**, *72*, 573–584. [CrossRef]
14. Ramakrishnan, S.; Ferrando, M.; Aceña-Muñoz, L.; De Lamo-Castellví, S.; Güell, C. Fish Oil Microcapsules from O/W Emulsions Produced by Premix Membrane Emulsification. *Food Bioprocess Technol.* **2013**, *6*, 3088–3101. [CrossRef]
15. Ramakrishnan, S.; Ferrando, M.; Aceña-Muñoz, L.; Mestres, M.; De Lamo-Castellví, S.; Güell, C. Influence of Emulsification Technique and Wall Composition on Physicochemical Properties and Oxidative Stability of Fish Oil Microcapsules Produced by Spray Drying. *Food Bioprocess Technol.* **2014**, *7*, 1959–1972. [CrossRef]
16. Tamm, F.; Gies, K.; Diekmann, S.; Serfert, Y.; Strunskus, T.; Brodkorb, A.; Drusch, S. Whey protein hydrolysates reduce autoxidation in microencapsulated long chain polyunsaturated fatty acids. *Eur. J. Lipid Sci. Technol.* **2015**, *117*, 1960–1970. [CrossRef]
17. Morales-Medina, R.; García-Moreno, P.J.; Pérez-Gálvez, R.; Muñío, M.; Guadix, A.; Guadix, E.M. Seasonal variations in the regiodistribution of oil extracted from small-spotted catshark and bogue. *Food Funct.* **2015**, *6*, 2646–2652. [CrossRef]
18. Camacho, F.; González-Tello, P.; Páez-Dueñas, M.P.; Guadix, E.M.; Guadix, A. Correlation of base consumption with the degree of hydrolysis in enzymic protein hydrolysis. *J. Dairy Res.* **2001**, *68*, 251–265. [CrossRef]
19. García-Moreno, P.J.; Pelayo, A.; Yu, S.; Busolo, M.; Lagaron, J.M.; Chronakis, I.S.; Jacobsen, C. Physicochemical characterization and oxidative stability of fish oil-loaded electrosprayed capsules: Combined use of whey protein and carbohydrates as wall materials. *J. Food Eng.* **2018**, *231*, 42–53. [CrossRef]

20. Danviriyakul, S.; McClements, D.J.; Decker, E.; Nawar, W.W.; Chinachoti, P. Physical stability of spray-dried milk fat emulsion as affected by emulsifiers and processing conditions. *J. Food Sci.* **2002**, *67*, 2183–2189. [CrossRef]
21. Bligh, E.G.; Dyer, W.J. A rapid method of total lipid extraction and purification. *Can. J. Biochem. Physiol.* **1959**, *37*, 911–917. [CrossRef]
22. Shantha, N.C.; Decker, E.A. Rapid, sensitive, iron-based spectrophotometric methods for determination of peroxide values of food lipids. *J. AOAC Int.* **1994**, *77*, 421–424. [CrossRef] [PubMed]
23. AOCS Official Method CE 8-89. *Determination of Tocopherols and Tocotrienols in Vegetable Oils and Fats by HPLC*; AOCS Press: Champaing, IL, USA, 1998.
24. Miguel, G.A.; Jacobsen, C.; Prieto, C.; Kempen, P.J.; Lagaron, J.M.; Chronakis, I.S.; García-Moreno, P.J. Oxidative stability and physical properties of mayonnaise fortified with zein electrosprayed capsules loaded with fish oil. *J. Food Eng.* **2019**, *263*, 348–358. [CrossRef]
25. Drusch, S.; Serfert, Y.; Berger, A.; Shaikh, M.Q.; Rätzke, K.; Zaporojtchenko, V.; Schwarz, K. New insights into the microencapsulation properties of sodium caseinate and hydrolyzed casein. *Food Hydrocoll.* **2012**, *27*, 332–338. [CrossRef]
26. ISO 6885:2006. *Animal and Vegetable Fats and Oils—Determination of Anisidine Value*; ISO: Geneva, Switzerland, 2006.
27. Hogan, S.A.; McNamee, B.F.; O'Riordan, E.D.; O'Sullivan, M. Emulsification and microencapsulation properties of sodium caseinate/carbohydrate blends. *Int. Dairy J.* **2001**, *11*, 137–144. [CrossRef]
28. Drusch, S.; Serfert, Y.; Scampicchio, M.; Schmidt-Hansberg, B.; Schwarz, K. Impact of physicochemical characteristics on the oxidative stability of fish oil microencapsulated by spray-drying. *J. Agric. Food Chem.* **2007**, *55*, 11044–11051. [CrossRef] [PubMed]
29. Sheu, T.Y.; Rosenberg, M. Microencapsulation by Spray Drying Ethyl Caprylate in Whey Protein and Carbohydrate Wall Systems. *J. Food Sci.* **1995**, *60*, 98–103. [CrossRef]
30. Young, S.L.; Sarda, X.; Rosenberg, M. Microencapsulating Properties of Whey Proteins. 2. Combination of Whey Proteins with Carbohydrates. *J. Dairy Sci.* **1993**, *76*, 2878–2885. [CrossRef]
31. Gharsallaoui, A.; Roudaut, G.; Chambin, O.; Voilley, A.; Saurel, R. Applications of spray-drying in microencapsulation of food ingredients: An overview. *Food Res. Int.* **2007**, *40*, 1107–1121. [CrossRef]
32. Jacobsen, C.; Rustad, T.; Nielsen, N.S.; Falch, E.; Jansson, S.; Storr, I. Processing of marine lipids and factors affecting their quality when used for functional foods. In *Marine Functional Food*; Luten, J.B., Ed.; Wageningen Academic Publishers: Wageningen, The Netherlands, 2009; Volume 2, pp. 89–114.
33. Jeyakumari, A.; Zynudheen, A.A.; Parvathy, U.; Binsi, P.K. Impact of chitosan and oregano extract on the physicochemical properties of microencapsulated fish oil stored at different temperature. *Int. J. Food Prop.* **2018**, *21*, 942–955. [CrossRef]
34. Unnikrishnan, P.; Puthenveetil Kizhakkethil, B.; Annamalai, J.; Ninan, G.; Aliyamveetil Abubacker, Z.; Chandragiri Nagarajarao, R. Tuna red meat hydrolysate as core and wall polymer for fish oil encapsulation: A comparative analysis. *J. Food Sci. Technol.* **2019**, *56*, 2134–2146. [CrossRef]
35. Drusch, S.; Rätzke, K.; Shaikh, M.Q.; Serfert, Y.; Steckel, H.; Scampicchio, M.; Voigt, I.; Schwarz, K.; Mannino, S. Differences in free volume elements of the carrier matrix affect the stability of microencapsulated lipophilic food ingredients. *Food Biophys.* **2009**, *4*, 42–48. [CrossRef]
36. Boerekamp, D.M.W.; Andersen, M.L.; Jacobsen, C.; Chronakis, I.S.; García-Moreno, P.J. Oxygen permeability and oxidative stability of fish oil-loaded electrosprayed capsules measured by Electron Spin Resonance: Effect of dextran and glucose syrup as main encapsulating materials. *Food Chem.* **2019**, *287*, 287–294. [CrossRef]
37. Drusch, S.; Berg, S. Extractable oil in microcapsules prepared by spray-drying: Localisation, determination and impact on oxidative stability. *Food Chem.* **2008**, *109*, 17–24. [CrossRef] [PubMed]
38. Frankel, E.N. Methods to determine extent of oxidation. In *Lipid Oxidation*; Woodhead Publishing Limited: Cambridge, UK, 2012; pp. 99–127.
39. Fu, X.; Xu, S.; Wang, Z. Kinetics of lipid oxidation and off-odor formation in silver carp mince: The effect of lipoxygenase and hemoglobin. *Food Res. Int.* **2009**, *42*, 85–90. [CrossRef]
40. Frankel, E.N. Foods. In *Lipid Oxidation*; Woodhead Publishing Limited: Cambridge, UK, 2012; pp. 299–354.

41. Drusch, S.; Serfert, Y.; Van Den Heuvel, A.; Schwarz, K. Physicochemical characterization and oxidative stability of fish oil encapsulated in an amorphous matrix containing trehalose. *Food Res. Int.* **2006**, *39*, 807–815. [CrossRef]
42. Elias, R.J.; McClements, D.J.; Decker, E.A. Antioxidant activity of cysteine, tryptophan, and methionine residues in continuous phase β-lactoglobulin in oil-in-water emulsions. *J. Agric. Food Chem.* **2005**, *53*, 10248–10253. [CrossRef]
43. Elias, R.J.; Kellerby, S.S.; Decker, E.A. Antioxidant activity of proteins and peptides. *Crit. Rev. Food Sci. Nutr.* **2008**, *48*, 430–441. [CrossRef]
44. Aluko, R.E. Amino acids, peptides, and proteins as antioxidants for food preservation. In *Handbook of Antioxidants for Food Preservation*; Woodhead Publishing: Cambridge, UK, 2015; pp. 105–140, ISBN 978-178-2420-972.
45. Hermund, D.; Jacobsen, C.; Chronakis, I.S.; Pelayo, A.; Yu, S.; Busolo, M.; Lagaron, J.M.; Jónsdóttir, R.; Kristinsson, H.G.; Akoh, C.C.; et al. Stabilization of Fish Oil-Loaded Electrosprayed Capsules with Seaweed and Commercial Natural Antioxidants: Effect on the Oxidative Stability of Capsule-Enriched Mayonnaise. *Eur. J. Lipid Sci. Technol.* **2019**, *121*, 1800396. [CrossRef]
46. Pal, R. Effect of Droplet Size on the Rheology of Emulsions. *AIChE J.* **1996**, *42*, 3181–3190. [CrossRef]
47. Schröder, A.; Berton-Carabin, C.; Venema, P.; Cornacchia, L. Interfacial properties of whey protein and whey protein hydrolysates and their influence on O/W emulsion stability. *Food Hydrocoll.* **2017**, *73*, 129–140. [CrossRef]
48. Yang, X.; Boyle, R.A. Sensory Evaluation of Oils/Fats and Oil/Fat-Based Foods. In *Oxidative Stability and Shelf Life of Foods Containing Oils and Fats*; Academic Press: Cambridge, MA, USA; AOCS Press: Urbana, IL, USA, 2016; pp. 157–185, ISBN 978-163-067-0566.
49. Let, M.B.; Jacobsen, C.; Meyer, A.S. Lipid oxidation in milk, yoghurt, and salad dressing enriched with neat fish oil or pre-emulsified fish oil. *J. Agric. Food Chem.* **2007**, *55*, 7802–7809. [CrossRef] [PubMed]
50. Nielsen, N.S.; Jacobsen, C. Retardation of Lipid Oxidation in Fish Oil-Enriched Fish Pâté-Combination Effects. *J. Food Biochem.* **2013**, *37*, 88–97. [CrossRef]
51. Sims, R.J.; Fioriti, J.A.; Trumbetas, J. Effect of sugars and sugar alcohols on autoxidation of safflower oil in emulsions. *J. Am. Oil Chem. Soc.* **1979**, *56*, 742–745. [CrossRef]

foods

MDPI

Article

Antiradical Activity of Hydrolysates and Extracts from Mollusk *A. broughtonii* and Practical Application to the Stabilization of Lipids

O.V. Tabakaeva [1,*], W. Piekoszewski [1,2,*], T.K. Kalenik [1], S.N. Maximova [3], A.V. Tabakaev [1], D. V. Poleshyk [3] and L. Proniewicz [4]

1 Department of Food Science and Technology, School of Biomedicine, Far Eastern Federal University, 690000 Vladivostok, Russia; kalenik.tk@dvfu.ru (T.K.K.); tabakaev.av@dvfu.ru (A.V.T.)
2 Department of Analytical Chemistry, Faculty of Chemistry, Jagiellonian University, 30-386 Krakow, Poland
3 Department of Food Technology, Far Eastern State Technical Fisheries University, 69000 Vladivostok, Russia; ipb_tpp@dgtru.ru (S.N.M.); tym1988@mail.ru (D.V.P.)
4 Department of Chemical Physics, Faculty of Chemistry, Jagiellonian University, 30-386 Krakow, Poland; leonard.proniewicz@uj.edu.pl
* Correspondence: tabakaeva.ov@dvfu.ru (O.V.T.); wojciech.piekoszewski@uj.edu.pl (W.P.)

Received: 29 December 2019; Accepted: 29 February 2020; Published: 7 March 2020

Abstract: The antiradical properties of hydrolysates and hydrothermal extracts of bivalve mollusks (*Anadara broughtonii*) from the Far Eastern Region of Russia and their influence on lipid oxidation in mayonnaise were investigated. The radical binding activity of hydrolysates and extracts of *A. broughtonii* varies from 55% to 89%. The maximum radical-binding activity was observed for acid hydrolysates. The antiradical efficiency of acid hydrolysates is 35%–41% of the BHT (butylhydroxytoluene) index. The antiradical activity depends on the (method of) technological and biotechnological processing of raw materials. Acid and enzymatic hydrolysates and hydrothermal extracts of *A. broughtonii* in mayonnaise slow down the process of oxidation of lipids and hydrolysis of triglycerides. Acid hydrolysates reduce the speed of oxidation and hydrolysis of lipids in mayonnaise more efficiently than the enzymatic hydrolysates.

Keywords: antiradical activity; hydrolysates; hydrothermal extracts; *A. broughtonii*

1. Introduction

The study of free-radical oxidation processes is important today due to their proven influence on various metabolic processes of the human body [1–3]. The accumulation of free radicals causes lipid peroxidation and dysfunction of the cell membrane, which leads to premature aging of the body, frequent sickness and even malignant tumors [4–6]. A quantitative study of the antiradical properties of given substances can suggest the ways in which such substances with antiradical activity are applied in practice. However, up till now, the focus of studies on antiradical activity has been mainly oriented towards vegetable raw materials—essential oils and mixtures thereof, and extracts of medicinal and other plants [7–9]. At the same time, raw materials of animal—including marine—origin, and products (of animal/marine origin) that have been subjected to technological and biotechnological modification could also have antioxidants. The biological and pharmacological activities of organic natural compounds of marine origin have been demonstrated by numerous studies [10–13].

Nowadays, many species of marine organisms that are used for food purposes, in traditional and non-traditional medicine, were also shown to have antioxidant activity and as raw materials for obtaining biologically active additives. Amongst marine hydrobionts, bivalve mollusks are promising sources of biologically active substances with antiradical properties. Substances from bivalves exhibit high antioxidant capacity: Extracts from *Crassula aequilatera*, *Mactra murchisoni*, and *Paphies donacina*, which were reported to scavenge 2,2-diphenyl-1-picrylhydrazyl (DPPH) radicals [14]; *Paphia malabarica* extracts, which were evaluated in relation to DPPH and $ABTS^+$ radical [15] crude extracts and fractions of alcohol extracts from *Mactra veneriformis* [16]; and oligopeptides from *Meretrix meretrix* [17].

The clams *Anadara broughtonii*, *Spisula sachalinensis*, *Mactra chinensis*, and others can be found in the coastal zone of the Sea of Japan [18]. The tissues of bivalve mollusks are rich in amino acids and specific carbohydrates [19–25]. *A. broughtonii* belongs to the family *Arcidae* and is a rather common type of Mollusca, in the *Bivalvia* class. It is found in the upper-sublittoral in subtropical areas of the Asian Pacific, with commercial stocks at depths of 2–15 m. The maximum lifespan of *A. broughtonii* is 65 years; the total mass of individuals varies from 80 to 380 g, and the shell length is 65–80 mm. *A. broughtonii* inhabits muddy and sandy bottoms, burrowing to depths of 10–25 cm. It is found in the Yellow Sea and the Sea of Japan [26].

Acid and enzymatic hydrolysates and hydrothermal extracts are products of technological and biotechnological processing of bivalve mollusks. They are sources of biologically active substances with potential health benefits [27,28].

Their antioxidant properties may be the basis for use as ingredients in lipid-containing emulsion products in order to reduce the rate of oxidation of lipids. Nowadays, various synthetic substances that perform the function of antioxidants are used for lipid stabilization of vegetable oils—potassium diethyl dithiophosphate or 2,4,6-tris-(dimethylamino) phenol or antioxidant mixtures containing BHT, naphthols, propyl gallates, and others. The disadvantages of these stabilizers are that they are not devoid of toxic properties [29]. Natural antioxidants are safer at stabilizing the quality of lipid rich foods [30].

The purpose of the present study was to investigate the antiradical properties of acid and enzymatic hydrolysates and hydrothermal extracts of the soft tissues of the bivalve mollusk *A. broughtonii* and to determine their effect on the oxidation of lipids in mayonnaise.

2. Materials and Methods

2.1. Material from the Bivalve Mollusk A. broughtonii

The clam *A. broughtonii* ranges from 65 to 80 mm in length, and its mass varies from 80 to 200 g. Molluscs were collected in June, September, and November 2016, and in February and April 2017 from Amur Bay (43060 N and 131440 E), Sea of Japan, in the Primorsky region in Russia (all analyses were done using pooled mollusks—seasonal changes were not taken into consideration in this study). The mollusks from all the sampling months were pooled to one sample of 5 kg. Live bivalve mollusks were transported under refrigeration (+6 °C) to the laboratory within 3 h and sampled randomly for this study. Upon arrival, the clams were manually shucked by cutting the adductor muscle with a knife. The clam juice was removed and the edible portion, constituting 12.46%–14.08% of the total weight of *A. broughtonii*, was collected. The edible portion was then dissected into 4 parts: Muscle, mantle, adductor, and viscera. Muscle and mantle were powdered using a blender (Phillips, Guangzhou, China) in the presence of liquid nitrogen. The samples were packed in a polyethylene bag, sealed and stored at −20 °C until use. The storage time was no longer than 1 month.

2.2. Acid and Enzymatic Hydrolysis and Hydrothermal Extraction and Analysis

Hydrolysates and extracts were obtained separately from the mantle and muscle of the mollusk. The frozen tissues of bivalve mollusks *A. broughtonii* were thawed at 4 °C before use, mixed with 6% solution of citric acid at a proportion of 1:1 according to the mass of tissues and acid, then homogenized in a tissue triturator for 2 min, and homogenate was obtained. The hydrolysis reaction was performed at 95 °C for 8 h. The resulting liquid phase was an acid hydrolysate (AH). In order to obtain enzymatic hydrolysate (EH), tissues of *A. broughtonii* were mixed with deionized water, the proportion being 1:2 according to the mass of tissues and water, then homogenized in a tissue triturator for 2 min, and homogenate was obtained. A hydrolysis reaction was performed at 45 °C and pH 7.5 in a shaking incubator at 150 rpm in order to promote activity of enzymes. The proteolytic enzyme Protomegaterin G20x (TU 00479942-002-94, Russia), produced by *Bacillus megaterium* was obtained from OOO PO SibBioPharm (Berdsk, Russia). Protomegaterin is a food-grade enzyme which has a declared proteolytic activity of 800 U/g. The enzyme was stored at 4 °C until it was used for the hydrolysis experiments.

The hydrolysis time was 8 h and the enzyme:substrate ratio was 5:1000. At the end of the hydrolysis, the liquid phase was separated by filtration, and boiled for 20 min in order to inactivate the enzyme. In order to obtain hydrothermal extracts (HTE), *A. broughtonii* tissues were mixed with deionized water in a 1:1 ratio according to the mass of tissues and water, and were then homogenized in a tissue triturator for 2 min, and homogenate was obtained. High-temperature processing was performed at 100 °C, duration 3 h. The moisture content was measured according to methodology described by the Association of Official Analytical Chemists (AOAC) (2000). Samples were dried in an oven at +105 °C until a constant weight was obtained [31].

2.3. Preparation of Mayonnaise

The mayonnaise ingredients included vegetable oil, egg powder, acid or enzymatic hydrolysate or hydrothermal extract, citric acid, mustard powder, salt, sugar, and water. The concentrations of additives (acid, enzymatic hydrolysates, hydrothermal extract) were determined on the basis of organoleptic indicators, primarily taste. The mayonnaise recipe is presented in Table 1.

Table 1. Composition of mayonnaise with hydrolysates and hydrothermal extracts from mollusks (*A. broughtonii*).

Component	Content, g/100 g		
	Mayonnaise with Acid Hydrolysate	Mayonnaise with Enzymatic Hydrolysate	Mayonnaise with Hydrothermal Hydrolysate
Vegetable oil	67	67	67
Egg powder	5	5	5
Acid hydrolysate	12	-	-
Enzymatic hydrolysate	-	18	-
Hydrothermal extract	-	-	20
Citric acid	-	0.4	0.4
Mustard powder	0.75	0.75	0.75
Salt	1.0	1.0	1.0
Sugar	1.5	1.5	1.5
Water	12.75	18.35	20.35
Total	100	100	100

Loose components (salt, sugar, citric acid, dry egg powder) were sieved and weighed out in accordance with recipes, then poured into food containers. Citric acid, sugar, and sodium chloride were dissolved in the required amount of water. In the case of using acid hydrolysates in the recipe, the procedure for the preparation of citric acid solution was excluded from the scheme of the technological process. The egg powder was mixed with water at a temperature of 40–45 °C and a ratio of 1:2, was heated to 60–65 °C, was held for 20–25 min for pasteurization, and then cooled to 30–40 °C. The acid hydrolysates or enzymatic hydrolysates or hydrothermal extracts, at a temperature of 55–60 °C, were mixed with the egg powder, mustard powder, and the sugar grains. The compound was stirred thoroughly and heated to 55–60 °C for 25–30 min in order to prepare the mayonnaise paste. After obtaining the mayonnaise paste and cooling it, an emulsion of mayonnaise was prepared. Vegetable oil at a temperature of 20 °C was injected into the mayonnaise paste while stirring. An acid-salt or brine solution was added after adding the whole oil dose, and the mixture was stirred for 15–20 min in order to further homogenize the emulsion. Final homogenization of the resulting mixture was carried out using a homogenizer (Kinematica, Polytron PT 45–80 GT, Switzerland). The pH value of mayonnaise was: 3.9 with acid hydrolysates, 4.1 with enzymatic hydrolysates, 4.3 with hydrothermal extracts, 3.9 with BHT, and 4.2 without additives. The samples were kept at room temperature (22 ± 2 °C). At designated times (30, 50, 70 and 90 days), the samples were taken for analyses. Mayonnaise without acid hydrolysates or enzymatic hydrolysates or hydrothermal extracts was used as a control. As a positive control, mayonnaise with the addition of BHT (Aldrich, 99.9%, analytical grade) was used.

2.4. Fractionation of Melanoidins

For fractionation of melanoidins, we used the method of gel-chromatography on columns with TSK-gels Toyopearl HW-40 and HW-50 ("Toyo Soda", Japan), pre-calibrated for proteins with well-known molecular weights. The gels that were used had the following ranges for separation of molecular masses for proteins: The separation of proteins with molecular masses from 100 to 10,000 Da takes place on HW-40; and on HW-50 for proteins from 500 to 80,000 Da. A 0.2 M solution of sodium chloride and distilled water was used as eluent. Samples hydrolysates and hydrothermal extracts were applied on the column (1.2 × 35 cm, free volume 12 mL) in 0.4 mL pre-skim chloroform.

The optical density of fractions of volume 1 mL was measured using a scanning spectrophotometer UV-1800 (Shimadzu, Japan) at 400 and 420 nm wavelengths. Calculation of the melanoidins content was carried out according to formulas derived on the basis of the standard curve [32].

2.5. DPPH Radical Scavenging Assay

Antiradical properties were evaluated in terms of ability to interact with the stable free radical 2,2-diphenyl-1-picrylhydrazyl (DPPH) in vitro. Determination was carried out in a reaction mixture containing 3 mL of 0.3 mM DPPH in ethanol, 1 mL 50 mM tris -HCl- buffer, pH 7.4, and 1 mL of extract or hydrolysate [33]. After 30 min incubation at a temperature of 20 °C, values of optical density were recorded at $\lambda = 517$ nm. The experiments were performed on a scanning spectrophotometer UV 1800 (Shimadzu, Japan) in cuvettes (l = 1 cm) at T = 298 °K

The activity was characterized by the following indicators:

- radical binding activity (RBA) was calculated by the formula

$$RBA\ (\%) = (D_{517}I - D_{517}II)/D_{517}I \times 100, \quad (1)$$

  where $D_{517}I$ is a control, and $D_{517}II$ is a sample;
- the effective concentration of substance at which 50% of free radicals *DPPH* (E_{C50}) was restored;
- time of recovery of half of the quantity of radical (T_{EC50}), min;
- antiradical efficiency (AE)—this characteristic connects the time of recovery of half of the quantity of radical (T_{EC50}) to the concentration of substrate (E_{C50}) necessary for this, which is calculated by the formula:

$$AE = 1/(E_{C50} \times T_{EC50}) \quad (2)$$

Antiradical properties were compared with the effect of the well-known synthetic antioxidant BHT (2,6-ditretbutyl-4-methyl-phenol), which was previously purified by recrystallization from ethanol, and then the isolated crystals were dried and sublimed in a vacuum.

2.6. Acid and Peroxide Value (AV, PV)

The acid and peroxide values (AV, PV) in the oil samples, in the oil samples extracted from mayonnaise with a mixture of solvents (ethyl ether and chloroform) were determined according to the methods stipulated in ISO [34,35]. The peroxide values were measured spectrophotometrically at 500 nm by a UV–1800 instrument (Shimadzu, Japan). Results were expressed in mmol of oxygen per kilogram of oil. Acid value (AV) is an important indicator of vegetable oil quality. AV is expressed as the amount of KOH (in milligrams) necessary to neutralize free fatty acids contained in 1 g of oil.

2.7. Statistical Analysis

Results were expressed as mean values (standard deviation) (n = 9). In total, three acid hydrolysates, three enzymatic hydrolysates, and three hydrothermal extracts were obtained. Data were subjected to analysis of variance (ANOVA), and mean comparisons were carried out using the Duncan's multiple range test (hydrolysates and hydrothermal extracts) and t-test (samples of mayonnaise with or without additives). Statistical analysis was performed using the statistical Package for Social Sciences (SPSS for windows: SPSS Inc., Chicago, IL, USA). Results with $p < 0.05$ were considered to be statistically significant.

3. Results and Discussion

The acid and enzymatic hydrolysates and hydrothermal extracts of motor muscle and mantle of *A. broughtonii* represent a dark brown liquid, with a specific faint odor. The content of dry substances was 15.9%–16.3% in acid hydrolysates, 8.3%–8.9% in enzymatic hydrolysates, and 5.6%–5.9% in hydrothermal extracts. The appearance of the received hydrolysates and extracts is shown in Figure 1.

(A) (B) (C)

Figure 1. Appearance of hydrolysates and extracts (**A**—acid hydrolysate (AH) muscle, **B**— enzymatic hydrolysates (EH) mantle, **C**—hydrothermal extracts (HTE) muscle).

Evaluation of the antiradical activity of the obtained acid and enzymatic hydrolysates and hydrothermal extracts of motor muscle and mantle of *A. broughtonii* and the well-known antioxidant BHT showed that all the studied objects have a high ability to bind radical DPPH (Table 2).

The radical-binding activity of acid and enzymatic hydrolysates and hydrothermal extracts of motor muscle and mantle of *A. broughtonii* varied within wide limits: From 55% to 89%. The results clearly indicate that the acid hydrolysate of mantle exhibited the highest radical binding activity (89%); this value is only 5% lower than for BHT. The minimum antiradical activity was shown by hydrothermal extracts of muscle, and this is characteristic for the extracts obtained from different parts of the bivalve mollusk *A. broughtonii*. Enzymatic hydrolysates possess lower antiradical activity than acid hydrolysates, but slightly higher than hydrothermal extracts. Our study showed that the method of modifying tissue of the bivalve mollusk *A. broughtonii* more significantly influences the level of manifestation of antiradical properties than the biological raw materials used. The data obtained are presented in Table 2.

Table 2. Antiradical activity of hydrolysates and extracts from mollusk *A. broughtonii*.

An Object	RBA%			EC_{50}, mkg/mL				T_{EC50}, min				AE, mkg/l·s × 10^{-2}			
	Subset for a = 0.05			Subset for a = 0.05				Subset for a = 0.05				Subset for a = 0.05			
	1	**2**	**3**	**1**	**2**	**3**	**4**	**1**	**2**	**3**	**4**	**1**	**2**	**3**	**4**
AH muscle	86.2 ± 3.21			16.1 ± 0.80				13.9 ± 0.65				0.45 ± 0.02			
AH mantle	89.3 + 4.02			15.5 ± 0.75				12.1 ± 0.60				0.53 ± 0.01			
EH muscle		70.5 ± 2.89				19.7 ± 0.94			14.5 ± 0.72				0.35 ± 0.01		
EH mantle		68.4 ± 2.95				22.6 ± 1.10			16.0 ± 0.80				0.28 ± 0.01		
HTE muscle			55.2 ± 2.17		27.5 ± 1.35					16.4 ± 0.80				0.22 ± 0.01	
HTE mantle			59.3 ± 2.58		30.2 ± 1.50					17.3 ± 0.81				0.19 ± 0.01	
BHT	94.3 ± 4.08						8.75 ± 0.41				7.00 ± 0.32				1.6 ± 0.07
Sig.	1.00	1.00	1.00	1.00	1.00	1.00		1.00	1.00	1.00		1.00	1.00	1.00	1.00

Means for groups in homogenous groups that are not statistically different are displayed. Data are mean ± standard deviation (n = 9).

The data in Table 2 demonstrate that acid hydrolysates have high antiradical efficiency (AE), and, furthermore, that acid hydrolysates from the mantle have higher values than those from muscle. The antiradical efficiency of acid hydrolysates is 35%–41% of the AE of BHT, which is a good indicator of antiradical properties. The acid hydrolysates of the mantle have 89% more AE than enzymatic hydrolysates, and 179% more than hydrothermal extracts. The AE of acid hydrolysates of muscle is 29% greater than that of enzymatic hydrolysates of muscle and 2.05 times greater than that of hydrothermal extracts (of muscle).

Similarly to our result, the protein hydrolysate obtained from the oyster *Saccostrea cucullata* exhibited 85.7% [10], the hydrolysate of the common smooth hound (*Mustelus mustelus*) exhibited 76.7% of scavenging activity [36], and protein hydrolysate of *Conus betulinus* exhibited 84% of scavenging activity [37]. By comparing the percentage of radical scavenging activity, the acid and enzymatic hydrolysates have proven to be a potential source for antiradical activity.

It is known that melanoidins and free amino acids exhibit antiradical properties. Previous research [38] has shown that acid hydrolysates of the tissues of mussels contain free amino acids, amines, dipeptides, free fatty acids, mineral substances, and high molecular substances, which are represented by two fractions: I—a molecular weight of 1500–6000 Da, and II—a weight less than 1500 Da. Moreover, the biological activity of fraction II was expressed significantly more strongly than fraction I. It was shown that fraction II comprises melanoidins having biological activity.

The fractional composition of high-molecular substances—acid and enzymatic hydrolysates and hydrothermal extracts—was researched in the present study. The quantitative proportions of fractions of melanoidins are presented in Figure 2.

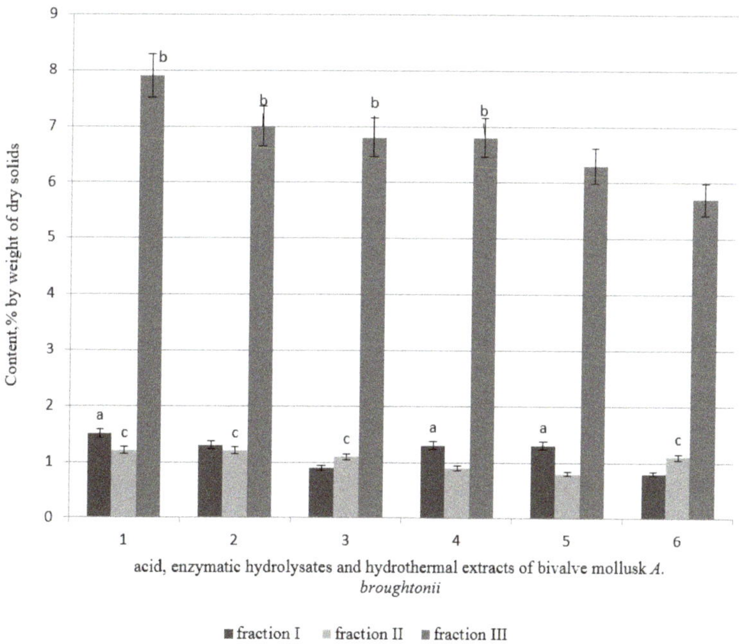

Figure 2. The proportion of melanoidins fractions in hydrolysates and extracts of bivalve mollusk *A. broughtonii* (1—AH muscle, 2—AH mantle, 3—enzymatic hydrolysate (EH) muscle, 4—EH mantle, 5—HTE muscle, 6—HTE mantle). Data are mean ± standard deviation (n = 9). letters indicate which fraction has the highest content of dry solids. a—fraction I versus fraction II ($p < 0.05$). b—fraction I versus fraction III ($p < 0.05$). c—fraction II versus fraction III ($p < 0.05$).

High molecular substances are divided into three fractions—two of them are substances with a molecular mass of about 10,000 Da, and the third fraction is of molecular weight of about 1000 Da. The maximum content of melanoidins was found in fraction III; fractions I and II contained only a small amount of melanoidins. The acid extract from the mantle was characterized by the highest content of melanoidins in fraction III.

Synthetic antioxidants such as butylated hydroxyanisole (BHA), butylated hydroxytoluene (BHT), tert-butylhydroquinone (TBHQ), and propyl gallate are often added to foods to slow lipid oxidation.

However, concerns have been raised about their safety as additives in food. Therefore, natural antioxidants with little or no side effects are of some interest as health-friendly supplements that can preserve food quality, primarily to protect fats from oxidation [39–41].

It is known that protein hydrolysates and peptides may exhibit antioxidant activity against lipid and fat peroxidation [39]. It has also been shown that hydrolysates and peptides of proteins of marine origin exhibit antioxidant activity in vitro and even have higher antioxidant activity than α-tocopherol [42–44].

Therefore, hydrolysates and hydrothermal extracts of bivalves can be used as a functional ingredient in the food industry, in particular in the production of mayonnaise.

Synthetic antioxidants such as butylated hydroxytoluene (BHT), butylated hydroxyanisole (BHA), and ethylenediaminetetraacetic acid (EDTA) (commercial antioxidants) are widely used in mayonnaise to prevent spoilage and the appearance of a bitter taste. However, these additives have a carcinogenic effect when used in high concentrations [45]. Consumers are demonstrating a growing demand for mayonnaise with natural ingredients. Adding natural antioxidants to food oil-fat emulsion systems, such as mayonnaise, has great potential to increase their oxidative stability and will be attractive to a wider group of consumers. Also, these compounds can have an impact on health, which will position mayonnaise as a healthy and natural product [46]. Gallic acid [47], ascorbic acid [48], tocopherol [49], rosemary [50], lactoferrin [51,52], phitic acid [52], mustard [53,54] lycopene [55], ginger powder [56], grape seed extract [57], essential oils extracted from Zenyan [58], chitosan [59], and seaweed [46] have been evaluated as natural antioxidants in mayonnaise.

However, protein hydrolysates are not used as antioxidants in the composition of mayonnaise. In the present study, acidic, enzymatic hydrolysates, and hydrothermal extracts in the composition of mayonnaise significantly inhibited lipid oxidation processes, as evidenced by a decrease in peroxide value compared to the control for all studied samples, but the value of the positive control was lower (Figure 3).

The minimum peroxide value (PV) during the storage period is determined for the fat released from the mayonnaise with acidic hydrolysates, which correlates well with the high antiradical properties of the hydrolysate established earlier. The enzymatic hydrolysates and hydrothermal extracts in mayonnaise also reduce the PV of mayonnaise lipids in proportion to their antiradical properties described above. Acid and enzymatic hydrolysates and hydrothermal extracts in mayonnaise reduce PV (90 days of storage) by 27%–50%. According to the data [54], lycopene crystals containing 50 mg/kg in mayonnaise reduced the peroxide number of mayonnaise by 54% during the three-month storage period. However, a ginger powder mass fraction in mayonnaise of 1% reduced the peroxide number by only 25% during 4 months of storage [55]. Thus, acid hydrolysates reduced the peroxide number as lycopene, hydrothermal extracts as ginger powder. Mayonnaise without additives had a shelf life of no more than 90 days; the peroxide number at the end of the shelf life exceeded 20 mmol O_2/kg. The shelf life of such mayonnaise was 17 days less than mayonnaise with additives of hydrolysates and hydrothermal extracts from the bivalve mollusk *A. broughtonii*.

The dynamics of changes of the acid value of fat isolated from mayonnaise with additives (acid, enzymatic hydrolysates and hydrothermal extracts and BHT) and without additives, during storage is presented in Figure 4.

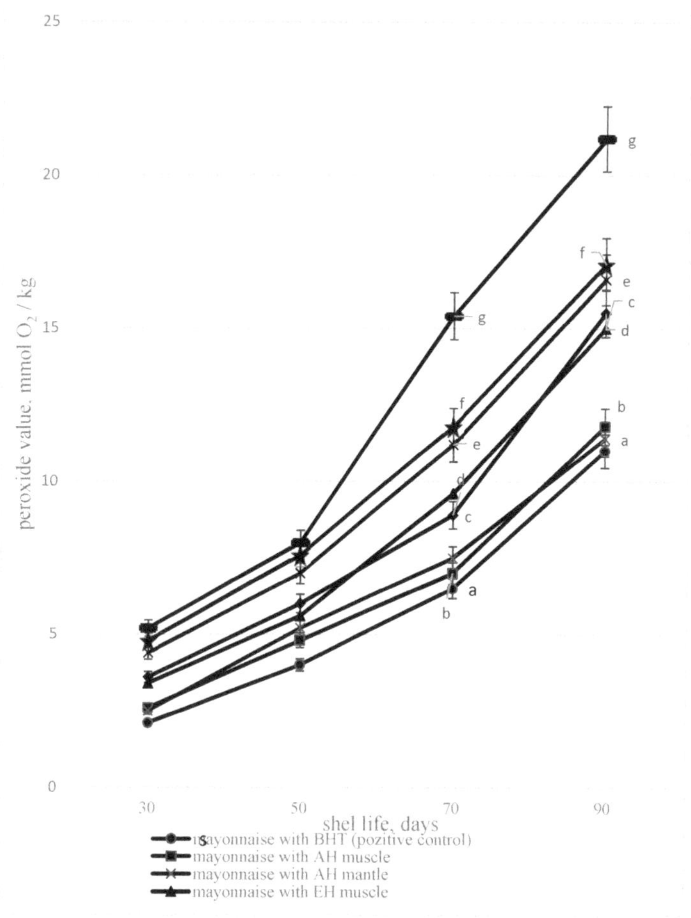

Figure 3. The peroxide value depending on the storage duration 1—mayonnaise with AH muscle, 2—mayonnaise with EH muscle, 3—mayonnaise with HTE muscle, 4—mayonnaise without additives (negative control), 5—mayonnaise with butylated hydroxytoluene (BHT) (positive control), 6—mayonnaise with AH mantle, 7—mayonnaise with EH mantle, 8—mayonnaise with HTE mantle). Data are mean ± standard deviation (n = 9). a—mayonnaise with AH muscle versus mayonnaise without additives (negative control); b—mayonnaise with AH mantle versus mayonnaise without additives (negative control); c—mayonnaise with EH mantle versus mayonnaise without additives (negative control); d—mayonnaise with EH muscle versus mayonnaise without additives (negative control), e—mayonnaise with HTE muscle versus mayonnaise without additives (negative control), f—mayonnaise with HTE mantle versus mayonnaise without additives (negative control), and g—mayonnaise with BHT (positive control) versus mayonnaise without additives (negative control) ($p < 0.05$).

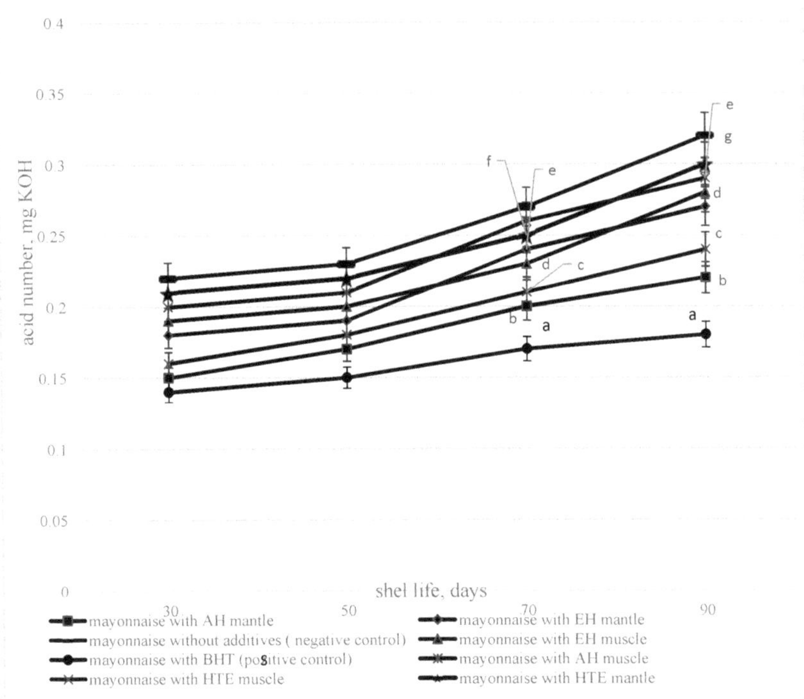

Figure 4. The acid value depending on the storage duration (1—mayonnaise with AH muscle, 2—mayonnaise with EH muscle, 3—mayonnaise with HTE muscle, 4—mayonnaise without additives (negative control), 5—mayonnaise with BHT (positive control), 6—mayonnaise with AH mantle, 7—mayonnaise with EH mantle, 8—mayonnaise with HTE mantle). Data are mean ± standard deviation (n = 9). a—mayonnaise with AH muscle versus mayonnaise without additives (negative control); b—mayonnaise with AH mantle versus mayonnaise without additives (negative control); c—mayonnaise with EH mantle versus mayonnaise without additives (negative control); d—mayonnaise with EH muscle versus mayonnaise without additives (negative control), e—mayonnaise with HTE muscle versus mayonnaise without additives (negative control), f—mayonnaise with HTE mantle versus mayonnaise without additives (negative control), and g—mayonnaise with BHT (positive control) versus mayonnaise without additives (negative control) ($p < 0.05$)

The obtained experimental data show that during storage—in addition to the oxidation of the fat phase of mayonnaise—hydrolysis of triglycerides to form free fatty acids is observed: The AV of oil, extracted from the mayonnaise, increases during storage. Acid and enzymatic hydrolysates and hydrothermal extracts in the composition of mayonnaise slow down triglyceride hydrolysis, as evidenced by the decrease of the AV of mayonnaise compared to the control for all investigated samples. The minimum AV throughout the storage period is determined for the fat extracted from the mayonnaise with acid hydrolysate. Acid and enzymatic hydrolysates and hydrothermal extracts in mayonnaise reduce the AV (at 90 days storage) by 20%–44%.

4. Conclusions

The hydrolysates and hydrothermal extracts of the bivalve mollusk *A. broughtonii* possess antiradical activity, the level of manifestation of which depends on the method of biotechnological and technological processing. Assessment of antiradical activity using the DPPH radical showed that acid hydrolysates have the maximum antiradical properties, probably due to the higher content of low molecular fractions of melanoidins.

Acid and enzymatic hydrolysates, and hydrothermal extracts in the mayonnaise composition slow down the processes of lipid oxidation and hydrolysis of triglycerides. Acid hydrolysates play the most significant role in reducing the speed of oxidation and hydrolysis. The results of this study can be practically applied in the production of mayonnaise and fat-containing sauces. However, the hydrolysates and extracts were not able to prevent the formation of peroxides as is the case for EDTA, which is a much better antioxidant in mayonnaise than BHT. Moreover, peroxides do not have any effect on off-flavors. To make a more complete evaluation of the effect of the hydrolysates and extracts determination of secondary lipid oxidation products such as aldehydes and ketones is necessary and preferably sensory analysis should also be done.

Author Contributions: O.V.T. and A.V.T. provided the idea of the research; and carried out the whole experiment (O.V.T. and A.V.T. contributed to this article equally); D.V.P. verified some of the experimental results; T.K.K. and W.P. analyzed and compared the results of the experiment; S.N.M. verified some of the experimental results; O.V.T., W.P., L.P. provided the framework of the paper and finally checked the quality of the article. All authors have read and agreed to the published version of the manuscript.

Funding: This research received no external funding.

Conflicts of Interest: The authors declare no conflict of interest.

References

1. Hancock, J.T.; Desikan, R.; Neil, S.J. Role of reactive oxygen species in cell signaling pathways. *Biochem. Soc. Trans.* **2001**, *29*, 345–350. [CrossRef] [PubMed]
2. Leanderson, P.; Faresjo, A.O.; Tagesson, C. Green tea polyphenols inhibits oxidant-induced DNA strand breakage in cultured lung cells. *Free Rad. Bio. Med.* **1997**, *23*, 235–242. [CrossRef]
3. Debashis, D.D.; Bhattacharjee, B.M.; Banerjee, R.K. Hydroxyl radicals are the major causative factor in stressinduced gastric ulceration. *Free Radic. Bio. Med.* **1997**, *23*, 8–18.
4. Rimm, E.B.; Ascherio, A.; Giovannucci, E.; Spiegelman, D.; Stampfer, M.J.; Willett, W.C. Vegetable, fruit, and cereal fiber intake and risk of coronary heart disease among men. *J. Am. Med. Assoc.* **1996**, *275*, 447–451. [CrossRef] [PubMed]
5. Steinmetz, K.A.; Potter, J.D. Vegetables, fruit, and cancer prevention: A review. *J. Am. Diet. Assoc.* **1996**, *96*, 1027–1039. [CrossRef]
6. Zenkov, N.K.; Lankin, V.Z.; Menshikova, E.B. *Okislitelnyi Stress*; Science: Moskow, Russia, 2001; p. 342.
7. Wettasinghe, M.; Shahidi, F. Evening primrose meal: A source of natural antioxidants and scavenger of hydrogen-derived free radicals. *J. Agric. Food Chem.* **1999**, *47*, 1801–1812. [CrossRef] [PubMed]
8. Ratz-Łyko, A.; Herman, A.; Arct, J.; Pytkowska, K. Evaluation of Antioxidant and Antimicrobial Activities of Oenothera biennis, Borago officinalis, and Nigella sativa Seedcake Extracts. *Food Sci. Biotechnol.* **2014**, *23*, 1029–1036. [CrossRef]
9. Zheng, W.; Wang, S.Y. Antioxidant activity and phenolic compounds in selected herbs. *J. Agric. Food Chem.* **2001**, *49*, 5165–5170. [CrossRef]
10. Umayaparvathi, S.; Arumugam, M.; Meenakshi, S.; Balasubramanian, T. Antioxidant Properties of Protein Hydrolysate Obtained from Oyster *Saccostrea cucullata*. *J. Aquat. Food Prod. Technol.* **2015**, *24*, 502–515. [CrossRef]
11. Rajapakse, N.; Mendis, E.; Byun, H.G.; Kim, S.K. Purification and in vitro antioxidative effects of giant squid muscle peptides on free radical-mediated oxidative systems. *J. Nutr. Biochem.* **2005**, *16*, 562–569. [CrossRef]
12. Thiansilakul, Y.; Benjakul, S.; Shahidi, F. Compositions, functional properties and antioxidative activity of protein hydrolysates prepared from round scad (*Decapterus maruadsi*). *Food Chem.* **2007**, *103*, 1385–1394. [CrossRef]
13. Cheorun, J.; Khan, F.F.; Khan, M.I.; Iqbal, J. Marine bioactive peptides: Types, structures, and physiological functions. *Food Rev. Int.* **2017**, *33*, 44–61.

14. Odeleye, T.; Li, Y.; Lindsey, W.W.; Nie, S.; Chen, S.; Wang, J.; Lu, J. The antioxidant potential of the New Zealand surf clams. *Food Chem.* **2016**, *204*, 141–149. [CrossRef] [PubMed]
15. Joy, M.; Chakraborty, K.; Raola, V.K. New sterols with anti-inflammatory potentials against cyclooxygenase-2 and 5-lipoxygenase from *Paphia malabarica*. *Nat. Prod. Res.* **2017**, *31*, 1286–1298. [CrossRef]
16. Luan, H.M.; Wang, L.C.; Wu, H.; Jin, Y.; Ji, J. Antioxidant activities and antioxidative components in the surf clam, *Mactra veneriformis*. *Nat. Prod. Res.* **2011**, *25*, 1838–1848. [CrossRef] [PubMed]
17. Huang, F.; Zhao, S.; Yu, F.; Yang, Z.; Ding, G. Protective effects and mechanism of *Meretrix meretrix* oligopeptides against nonalcoholic fatty liver disease. *Mar. Drugs* **2017**, *15*, 31. [CrossRef]
18. Arzamastsev, I.S.; Yakovlev Yu, M.; Evseev, G.A. *Atlas of Trade Invertebrates and Seaweed of the Seas of the Far East of Russia*; Avante: Vladivostok, Russia, 2001; p. 192.
19. De-Wei, C.; Jian, S.; Xiao-Ling, L.; Dong-Mei, Y.; Ying, L.; Wei-Ming, J.; Xiao-Han, C. Amino Acid Profiles of Bivalve Mollusks from Beibu Gulf, China. *J. Aquat. Food Prod. Technol.* **2012**, *21*, 369–379.
20. Karnjanapratum, S.; Benjakul, S.; Kishimura, H.; Tsai, Y.H. Chemical compositions and nutritional value of Asian hard clam (Meretrixlusoria) from the coast of Andaman Sea. *Food Chem.* **2013**, *141*, 4138–4145. [CrossRef]
21. Tabakaeva, O.V.; Tabakaev, A.V.; Pikozevski, V. Nutritional composition and total collagen content of two commercially important edible bivalve molluscs from the Sea of Japan coast. *J. Food Sci. Technol.* **2018**, *55*, 4877–4886. [CrossRef]
22. Tabakaeva, O.V.; Tabakaev, A.V. Amino acids composition of soft tissues of the Far-east bivalve mollusk *Anadara broughtonii*. *Chem. Nat. Compd.* **2016**, *52*, 468–471. [CrossRef]
23. Wright, A.C.; Fan, Y.; Baker, G.L. Nutritional value and food safety of bivalve molluscan shellfish. *J. Shellfish. Res.* **2018**, *37*, 695–708. [CrossRef]
24. Tabakaeva, O.V.; Tabakaev, A.V. Amino-acid profile of a mactridae bivalve mollusk from the sea of Japan. *Chem. Nat. Compd.* **2016**, *52*, 966–968. [CrossRef]
25. Minju, J.; Kajal, C. Nutritional Qualities of the Low-Value Bivalve Mollusks Paphia malabarica and villorita cyprinoides at the Estuarine Waters of the Southwestern Coast of India. *J. Aquat. Food Prod. Technol.* **2017**, *26*, 54–70.
26. Rakov, V.A. *Determinants of Two-Fold Molluscs of Primorsky Krai*; Dalnauka: Vladivostok, Russia, 2002.
27. Yang, X.; Qiu, Y.; Zhao, Y.; Chi, C.; Wang, B. Purification and Characterization of Antioxidant Peptides Derived from Protein Hydrolysate of the Marine Bivalve Mollusk *Tergillarca granosa*. *Mar. Drugs* **2019**, *17*, 251. [CrossRef]
28. Qiao, M.; Tu, M.; Chen, H.; Mao, F.; Yu, C.; Du, M. Identification and in Silico Prediction of Anticoagulant Peptides from the Enzymatic Hydrolysates of Mytilus edulis Proteins. *Int. J. Mol. Sci.* **2018**, *19*, 2100. [CrossRef]
29. Eshghi, N.; Asnaashari, M.; Haddad Khodaparast, M.H.; Hosseini, F. Evaluating the potential of natural curcumin for oxidative stability of soybean oil. *Nat. Prod. Res.* **2014**, *28*, 1375–1378. [CrossRef]
30. Shadyro, O.I.; Sosnovskaya, A.A.; Edimecheva, I.P. Flaxseed oil stabilization using natural and synthetic antioxidants. *Eur. J. Lipid Sci. Technol.* **2017**, *119*, 1700079. [CrossRef]
31. AOAC. *Official Method of Analytical Chemists*, 17th ed.; Association of Official Analytical Chemists: Washington, DC, USA, 2000.
32. Brands, C.M.; Wedzicha, B.L.; Boekel, M.A. Quantification of melanoidin concentration in sugar-casein systems. *J. Agric. Food Chem.* **2002**, *50*, 1178–1183. [CrossRef]
33. Brand-Williams, W.; Cuvelier, M.E.; Berset, C. Use of a free radical method to evaluate antioxidant activity. *LWT Food Sci. Technol.* **1995**, *28*, 25–30. [CrossRef]
34. ISO 660, Animal and Vegetable Fats and Oils. *Determination of Acid Value and Acidity*; International Standards Organization: Geneva, Switzerland, 2009.
35. ISO 3960, Animal and Vegetable Fats and Oils. *Determination of Peroxide Value-Lodometric (Visual) Endpoint Determination*; International Standards Organization: Geneva, Switzerland, 2007.
36. Bougatef, A.; Hajji, M.; Balti, R.; Lassoued, I.; Triki-Ellouz, Y.; Nasri, M. Antioxidant and free radical-scavenging activities of smooth hound (*Mustelus mustelus*) muscle protein hydrolysates obtained by gastrointestinal proteases. *Food Chem.* **2009**, *114*, 1198–1205. [CrossRef]
37. Nazeer, R.A.; Srividhya, T.S. Antioxidant peptides from the protein hydrolysates of *Conus betulinus*. *Int. J. Pept. Res. Ther.* **2011**, *17*, 231–237. [CrossRef]

38. Terentyev, V.A. Development of the production technology of the media hydrolyzate of treatment-and-prophylactic application. *Xranenie Pererab. Selxozsirya* **2002**, *4*, 12–14.
39. Harnedy Padraigin, A.; FitzGerald Richard, J. Bioactive peptides from marine processing waste and shellfish. *J. Funct. Foods* **2012**, *4*, 6–24. [CrossRef]
40. Chirag, R.; Suchita, G.; Pritam, K. Antioxidant obtained from natural sourse. *Indo Am. J. Pharm. Sci.* **2019**, *6*, 4773–4780.
41. Giulia, M.; Dania, G. Antibacterial and antioxidant activities for natural and synthetic dual-active compounds. *Eur. J. Med. Chem.* **2018**, *158*, 91–105.
42. Kim, S.K.; Lee, H.C.; Byun, H.G.; Jeon, Y.J. Isolation and characterization of antioxidative peptides from enzymatic hydrolysates of yellowfin sole skin gelatin. *J. Korean Fish. Soc.* **1996**, *41*, 198–204.
43. Cho, W.I.; Kim, S.M. Taste compounds and biofunctional activities of the sandy beach clam hydrolysate for the shellfish flavoring condiment. *J. Aquat. Food Prod. Technol.* **2016**, *25*, 24–34. [CrossRef]
44. Lorenzo, J.M.; Munekata, P.E.; Gómez, B.; Barba, F.J.; Mora, L.; Pérez-Santaescolástica, C.; Toldrá, F. Bioactive peptides as natural antioxidants in food products. *Trends Food Sci. Technol.* **2018**, *79*, 136–147. [CrossRef]
45. Magdalena, M.; Antonia, M.M. Antioxidant Activity of Resveratrol Compared with Common Food Additives. *J. Food Prot.* **2001**, *64*, 379–384.
46. Hermund, D.B.; Yesiltas, B.; Honold, P.; Jonsdottir, R.; Kristinsson, H.G.; Jacobsen, C. Characterisation and antioxidant evaluation of Icelandic F. vesiculosus extracts in vitro and in fish-oil-enriched milk and mayonnaise. *J. Funct. Foods* **2015**, *19*, 828–841. [CrossRef]
47. Jacobsen, C.; Adler-Nissen, J.; Meyer, A.S. Effect of ascorbic acid on iron release from the emulsifier interface and on the oxidative flavor deterioration in fish oil enriched mayonnaise. *J. Agric. Food Chem.* **1999**, *47*, 4917–4926. [CrossRef] [PubMed]
48. Jacobsen, C.; Timm, M.; Meyer, A.S. Oxidation in fish oil enriched mayonnaise: Ascorbic acid and low pH increase oxidative deterioration. *J. Agric. Food Chem.* **2001**, *49*, 3947–3956. [CrossRef] [PubMed]
49. Sørensen, A.D.M.; Nielsen, N.S.; Hyldig, G.; Jacobsen, C. Influence of emulsifier type on lipid oxidation in fish oil-enriched light mayonnaise. *Eur. J. Lipid Sci. Technol.* **2010**, *112*, 1012–1023. [CrossRef]
50. Jacobsen, C.; Xu, X.; Skall Nielsen, N.; Timm-Heinrich, M. Oxidative stability of mayonnaise containing structured lipids produced from sunflower oil and caprylic acid. *Eur. J. Lipid Sci. Technol.* **2003**, *105*, 449–458. [CrossRef]
51. Nielsen, N.S.; Petersen, A.; Meyer, A.S.; Timm-Heinrich, M.; Jacobsen, C. Effects of lactoferrin, phytic acid, and EDTA on oxidation in two food emulsions enriched with long-chain polyunsaturated fatty acids. *J. Agric. Food Chem.* **2004**, *52*, 7690–7699. [CrossRef]
52. Lagunes-Galvez, L.; Cuvelier, M.E.; Odonnaud, C.; Berset, C. Oxidative stability of some mayonnaise formulations during storage and daylight irradiation. *J. Food Lipids* **2002**, *9*, 211–224. [CrossRef]
53. Milani, M.A.; Mizani, M.; Ghavami, M.; Eshratabadi, P. The physicochemical influences of yellow mustard paste-comparison with the powder in mayonnaise. *J. Food Process. Technol.* **2013**, *4*, 2. [CrossRef]
54. Kaur, D.; Wani, A.A.; Singh, D.P.; Sogi, D. Shelf life enhancement of butter, ice-cream, and mayonnaise by addition of lycopene. *Int. J. Food Prop.* **2011**, *14*, 1217–1231. [CrossRef]
55. Kishk, Y.; Elsheshetawy, H.E. Effect of ginger powder on the mayonnaise oxidative stability, rheological measurements, and sensory characteristics. *Ann. Agric. Sci.* **2013**, *58*, 213–220. [CrossRef]
56. Altunkaya, A.; Hedegaard, R.V.; Harholt, J.; Brimer, L.; G€okmen, V.; Skibsted, L.H. Oxidative stability and chemical safety of mayonnaise enriched with grape seed extract. *Food Funct.* **2013**, *4*, 1647–1653. [CrossRef]
57. Hashemi, M.B.; Niakousari, M.; Saharkhiz, M.J. Antioxidant activity of Satureja bachtiarica Bunge essential oil in rapeseed oil irradiated with UV rays. *Eur. J. Lipid Sci. Technol.* **2011**, *113*, 1132–1137. [CrossRef]
58. Baranauskiene, R.; Kazernaviciute, R.; Pukalskiene, M.; Mazdzieriene, R.; Venskutonis, P.R. Agrorefinery of Tanacetum vulgare L. into valuable products and evaluation of their antioxidant properties and phytochemical composition. *Ind. Crop. Prod.* **2014**, *60*, 113–122. [CrossRef]
59. Honold, P.J.; Jacobsen, C.; Jonsdottir, R.; Kristinsson, H.G.; Hermund, D.B. Potential seaweed-based food ingredients to inhibit lipid oxidation in fish-oilenriched mayonnaise. *Eur. Food Res. Technol.* **2016**, *242*, 571–584. [CrossRef]

foods

MDPI

Article

Anti-Hyperglycemic Effects of Green Crab Hydrolysates Derived by Commercially Available Enzymes

Bouhee Kang [1], Denise I. Skonberg [1,*] and Angela D. Myracle [2]

[1] School of Food and Agriculture, University of Maine, 5735 Hitchner Hall, Orono, ME 04469, USA; bouhee.kang@maine.edu

[2] School of Food and Agriculture, University of Maine, 5763 Rogers Hall, Orono, ME 04469, USA; angela.myracle@maine.edu

* Correspondence: denise.skonberg@maine.edu; Tel.: +1-207-581-1639

Received: 15 January 2020; Accepted: 25 February 2020; Published: 28 February 2020

Abstract: The predation and burrowing activity of invasive green crabs have had detrimental effects on important marine resources and habitats. Our objective is to develop bioactive hydrolysates by enzymatic proteolysis of underutilized green crab. Mechanically separated mince was hydrolyzed with Alcalase, Protamex, Flavourzyme, and Papain (1%) for 60 min. Subsequently, the hydrolysates were introduced to a simulated gastrointestinal digestion model. Selected samples were fractionated by ultrafiltration, and their anti-hyperglycemic effects including α-glucosidase, α-amylase, and dipeptidyl peptidase-IV (DPP-IV) inhibitory activities and glucagon-like 1 (GLP-1) secretory activity were evaluated. The Protamex treatment showed the highest α-glucosidase inhibitory activity (IC_{50} 1.38 ± 0.19 mg/mL) compared to other enzyme treatments and the crab mince control, and its α-amylase inhibitory activity (IC_{50} 11.02 ± 0.69 mg/mL) was lower than its α-glucosidase inhibitory activity. Its GLP-1 secretory activity was approximately four times higher than the positive control (10 mM glutamine). The <3 kD fraction contributed significantly to the anti-hyperglycemic activity of Protamex-derived hydrolysates, and this activity was stable after simulated digestion. Our results suggest that green crab hydrolysates obtained by Protamex treatment have the potential for type 2 diabetes management and could be incorporated in food products as a health-promoting ingredient.

Keywords: green crab; protein hydrolysates; enzymatic hydrolysis; type 2 diabetes; anti-hyperglycemia

1. Introduction

In recent years, consumer interest in sustainability and healthy lifestyles has significantly increased, leading to higher demand for nutritious and environmentally-friendly food products among consumers and the food industry. The development of natural, bioactive food ingredients from currently underutilized food resources may positively contribute to satisfying this demand. Nutritional and bioactive food ingredients include proteins and peptides, polyunsaturated fatty acids (PUFAs), phytochemicals, fibers, probiotics, and prebiotics [1]. Marine organisms represent good sources for obtaining various bioactive compounds, and many of these bioactive compounds are derived from proteins [2]. Therefore, protein-rich marine organisms are an ideal raw starting material for the development of protein-derived bioactive compounds.

European green crabs (*Carcinus maenas*) are an invasive species well established in the U.S. that negatively influence fisheries, aquaculture, and marine habitats due to their high resilience, voracious predation, and strong burrowing activity [3–6]. Green crabs are very competitive predators that prefer juvenile clams, blue mussels, and lobster larvae, which are highly profitable for fisheries and aquaculture, as their prey [3,4]. Their intensive burrowing and high viability have destroyed not only valuable eelgrass beds and salt marshes but also the overall balance of the natural marine ecosystem [5]. Although there is a large population of green crabs in the U.S., they are not currently utilized by food industries due to their small size. To obtain the nutritious meat, significant labor and high costs are required to separate the muscle from the carapace. However, our previous studies demonstrated that green crab mince obtained via mechanical separation can be used in the development of value-added food products [7,8].

Green crab contains various nutritional and nutraceutical compounds such as chitin, carotenoids, essential amino acids, and PUFAs including eicosapentaenoic acid (EPA) and docosahexaenoic acid (DHA) [9]. Green crab meat has approximately 80% protein [10]; these proteins have the potential to be utilized as the initial material for the generation of multifunctional food ingredients. Some proteins and peptides extracted from protein-rich marine resources have been reported to act as functional and bioactive ingredients in food products [11,12]. Enzymatic hydrolysis of a variety of protein sources has been extensively utilized as a method to generate bioactive peptides having antioxidant, anti-inflammatory, and anti-hypertensive effects [13–15]. Different types of proteases used for enzymatic hydrolysis have distinct specificity and produce peptides having different molecular weights and amino acid sequences, which may contribute to their diverse biological and physiochemical functional properties [16–18]. Various studies have confirmed that commercially available proteases such as Alcalase, Protamex, Flavourzyme, and Papain have the potential to generate bioactive peptides from food proteins [19,20]. However, the majority of these studies were conducted with dairy proteins and finfish byproducts. Furthermore, the bioactivity of hydrolysates obtained from marine resources was studied extensively with regard to their anti-hypertensive and antioxidant activities rather than their anti-diabetic properties. Fewer studies have been conducted to assess the biofunctional activity of the valuable peptides derived from shellfish proteins. More specifically, there is a lack of research about the potential anti-diabetic activity of crustacean and mollusk proteins and their hydrolysates.

Type 2 diabetes mellitus (T2DM) is one of the most prevalent chronic diseases in the world and ranks in the top 10 diseases causing the most deaths worldwide (WHO) [21]. T2DM and associated complications result not only in health issues but also substantial direct and indirect costs [22]. However, an optimal diet can help in the prevention of T2DM and is critical to reducing the attendant morbidity and cost associated with the treatment of T2DM [23–25]. The development of anti-hyperglycemic ingredients from green crabs may contribute to the prevention of T2DM and reduce the use of pharmaceutical products and the risks associated with their side-effects [26]. Furthermore, the development of natural and nutraceutical peptides generated from green crabs may encourage the development of a fishery for this invasive predator. Therefore, the aim of our study was to investigate the anti-hyperglycemic effect of green crab hydrolysates for potential use as a health-promoting food ingredient. To achieve this, (1) enzymatic hydrolysis using commercially available proteases was applied to mechanically separated green crab mince, (2) the potential anti-hyperglycemic effects of green crab hydrolysates were investigated, (3) fractionation of hydrolysates using ultrafiltration was conducted, and (4) changes in anti-hyperglycemic activity of green crab hydrolysates after simulated gastrointestinal digestion were investigated.

2. Materials and Methods

2.1. Materials and Reagents

Green crabs (carapace width: 40–85 mm and weight: 20–155 g) were harvested in 2018 on the Back River in Georgetown, Maine, USA. All of the chemicals and reagents that were used in this study were supplied by Fisher Scientific (Hampton, NH, USA), Thermo Fisher Scientific (Waltham, MA, USA), Bio-Rad Laboratories (Hercules, CA, USA), and Sigma-Aldrich (St. Louis, MO, USA) unless otherwise described. Alcalase 2.4 L (AL), Protamex (PR), Flavourzyme (FL), and Papain (PA) were purchased from Sigma-Aldrich (St. Louis, MO, USA). Rat intestinal acetone powders of α-glucosidase (EC 3.2.1.20), porcine pancreatic α-amylase (EC 3.2.1.1), pepsin from porcine gastric mucosa (EC 3.4.23.1, >250 units/mg protein), pancreatin from porcine pancreas (EC 232.468.9, 8 × USP), p-nitrophenyl-β-D-glucopyranoside (p-NPG), and Gly-Pro-p-nitroanilide were purchased from Sigma-Aldrich (St. Louis, MO, USA). Porcine kidney dipeptidyl peptidase-IV (DPP-IV) (E.C. 3.4.14.5) and glucagon-like 1 (GLP-1) ELISA kit were purchased from EMD Millipore (Billerica, MA, USA). Cell culture materials including penicillin-streptomycin (0.1g/L), Dulbecco's modified Eagle's medium (DMEM) containing low glucose, Glutamax, trypsin/EDTA (10×), and Poly D-lysine were obtained from Gibco Life Technologies (Grand Island, NY, USA). Fetal bovine serum (FBS) was purchased from Atlanta Biologicals (Minneapolis, MN, USA). Bovine serum albumin, acarbose, sitagliptin, and glutamine were obtained from Sigma-Aldrich (St. Louis, MO, USA).

2.2. Preparation of Green Crab

Forty kilograms of harvested crabs were washed with tap water and then blast-frozen at −30 °C for 1 h. Frozen crabs were stored at −20 °C until further use. Partially thawed green crabs were processed through a mechanical separator (Paoli One-Step mechanical separator, Rockford, IL) to separate meat and shell streams, and then the minced meat was vacuum sealed in plastic bags (3 mil, 3.3 cm^3/100 in^2 oxygen transmission rate, 80 microns, UltraSource, Kansas, MO, USA). The crab mince was blast-frozen at −30 °C for 1 h and then stored at 20 °C until further use (Figure 1).

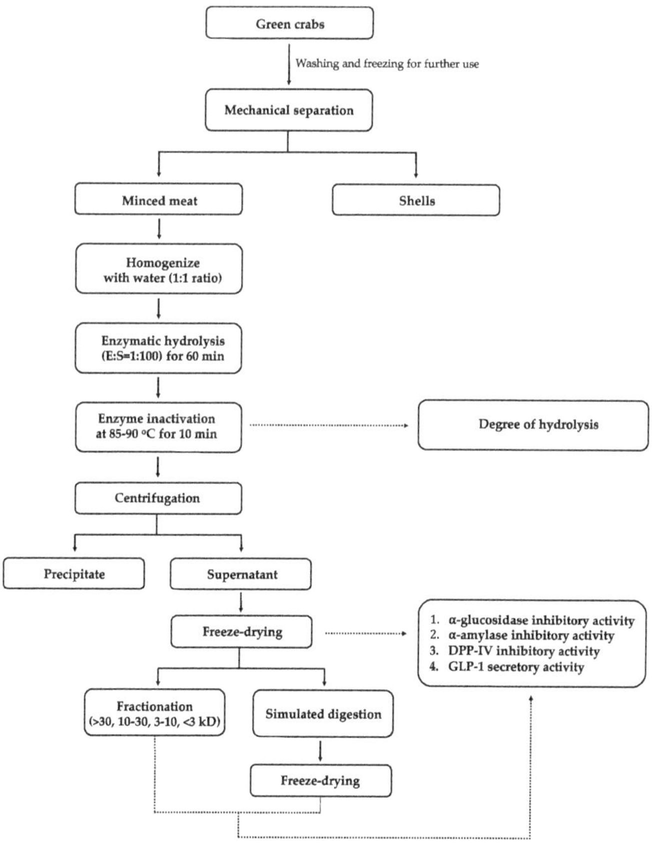

Figure 1. Process flow of sample preparation.

2.3. Enzymatic Hydrolysis of Crab Mince

Enzymatic hydrolysis using commercially available proteases was conducted based on a modified protocol from Beaulieu et al. [27]. The raw crab mince was mixed with deionized water in a 1:1 (*w/w*) ratio, then homogenized for 1 min in a Waring blender at maximum speed. The enzymatic hydrolysis was conducted at the optimum pH and temperature of each enzyme including Alcalase (AL, 50 °C and pH 8), Protamex (PR, 50 °C and pH 7), Flavourzyme (FL, 50 °C and pH 7), and Papain (PA, 65 °C and pH 6). Protein content was determined based on total nitrogen content as analyzed by a combustion analyzer (TRU MAC CNS, LECO Corp., St. Joseph, MI, USA). Protein content was determined using a protein conversion factor of 6.25. Enzyme was added to the homogenate based on substrate (protein) content (1:100 = E:S), and the mixture was hydrolyzed for 60 min. During hydrolysis, the pH of the mixture was maintained using 6 N HCl or 6 N NaOH and temperature was monitored. After hydrolysis, the mixture was heated at 85–90 °C for 10 min to inactivate the enzyme and then immediately cooled on ice. For the crab mince control (CMC), mince was homogenized with water but not subsequently treated with the enzyme. Subsampling was performed to evaluate the degree of hydrolysis, and then the mixtures were centrifuged at 19,722× *g* for 15 min at 4 °C, and the supernatants were collected. All of the treatments were processed in triplicate. The collected supernatants were blast-frozen at −30 °C for 1 h, then freeze-dried (35 EL, VirTis Co. Inc., Gardiner, NY, USA) at −30 to 25 °C under 250 mT for 10 days All lyophilized supernatants were stored at −80 °C until further use.

2.4. Degree of Hydrolysis

Degree of hydrolysis was determined following the O-phthalaldehyde (OPA) method [28,29]. OPA reagent was prepared with 375 mL of deionized water, 19.05 g of sodium tetraborate decahydrate, 500 mg of sodium dodecyl sulfate (SDS), and 400 mg of 97% OPA in 10 mL of ethanol. After mixing, 440 mg of 99% dithiothreitol (DTT) was added to the solution and deionized water was added to achieve a final volume of 500 mL. For the sample preparation, the CMC and enzyme hydrolysates were diluted with 4% SDS (1:19 *w/v*). After centrifugation at 1100× *g* for 10 min, 4 mL of supernatant was collected. Subsequently, the supernatant was diluted to 50 mL with deionized water. Four mL of OPA reagent were mixed with 400 μL of solubilized sample/standard (0.5 mg/mL serine) and the mixture was incubated at room temperature for 2 min. Absorbance was measured at 340 nm and the degree of hydrolysis was calculated based on the following three equations:

$$\text{Serine-NH}_2\ (\text{mequiv/g protein}) = \frac{\frac{(\text{OD}_{\text{sample}} - \text{OD}_{\text{blank}})}{(\text{OD}_{\text{standard}} - \text{OD}_{\text{blank}})} \times 0.9516 \frac{\text{mequiv}}{\text{L}} \times 0.05 \times 100}{\text{W} \times \frac{\text{V1}}{\text{V2}} \times \text{P}} \tag{1}$$

W = weight of sample (g);
P = protein (g/100 g sample);
V1 = volume of supernatant (4 mL);
V2 = volume of SDS solution (20 mL);
0.05 = sample volume in L.

$$\text{h} = \frac{(\text{Serine} - \text{NH2}) - \beta}{\alpha} \tag{2}$$

h = number of hydrolyzed peptide bonds;
α = 1.00, β = 0.40 mequiv/g protein;

$$\text{DH} = \frac{\text{h}}{\text{h}_{\text{total}}} \times 100 \tag{3}$$

Total number of peptide bonds (h_{total}) of fish: 8.6 mequiv/g protein.

2.5. Simulated Gastrointestinal Digestion

Simulated gastrointestinal digestion of the control and crab mince hydrolysates was performed according to González-Montoya et al. [30]. The sample was dissolved in deionized water (5% *w*/*v*), then pH was adjusted to pH 2 using 6 N HCl prior to heating to 37 °C. Pepsin was added to the sample mixture (4:100 = E:S protein/peptide content) and stirred at 37 °C for 60 min. Then, pH was adjusted to 7.5 using 6 N NaOH. Subsequently, pancreatin (4:100 = E:S protein/peptide content) was added for the intestinal digestion phase and the mixture was stirred at 37 °C for 2 h. To inactivate the digestive enzymes, the mixture was heated at 85–90 °C for 10 min. The mixture was freeze-dried (35 EL, VirTis Co. Inc., Gardiner, NY, USA) at −30 to 25 °C under 250 mT for 10 days and then stored at −80 °C until further use.

2.6. Sodium Dodecyl Sulfate Polyacrylamide Gel Electrophoresis (SDS-PAGE)

Molecular weight distribution was determined by SDS-PAGE. All of the samples were dissolved in a Laemmli sample buffer containing 62.5 mM Tris-HCl (pH 6.8), 2% SDS, 25% (*v*/*v*) glycerol, 0.01% bromophenol blue, and 5% β-mercaptoethanol, then heated in a 90–95 °C water bath for 5 min [31]. Samples were separated on a 4% stacking and 16.5% separating gel (Bio-Rad Laboratories, Hercules, CA) using Tris/Glycine/SDS running buffer (25 mM Tris, 192 mM glycine, 0.1% (*w*/*v*) SDS) in a Bio-Rad Mini-PROTEAN III Cell (Bio-Rad Laboratories, Inc., Hercules, CA, USA). The amount that was loaded to the wells was adjusted to 1 mg for the hydrolysates, 0.5 mg for the control, and 10 µL of a protein molecular weight standard (Precision Plus Protein™ Dual Xtra Prestained Protein Standards 2-250 kD, Bio-Rad Laboratories, Inc., Hercules, CA, USA). Separated proteins were fixed in a gel fixing solution containing 50% (*v*/*v*) ethanol and 10% (*v*/*v*) acetic acid then washed with a solution consisting of 50% (*v*/*v*) methanol and 10% (*v*/*v*) acetic acid. Following washing, gel staining was performed using Coomassie blue R-250 solution at room temperature for 3 h with gentle agitation. (Bio-Rad Laboratories, Inc., Hertfordshire, UK), then the gel was destained with washing solution until the background color was removed.

2.7. Fractionation Using Ultrafiltration

Green crab hydrolysates were fractionated using ultrafiltration based on molecular weight. A volume of 15 mL of solubilized sample (50 mg/mL) was added into a 30 kD molecular weight cut-off (MWCO) filter device (Amicon® ultra-15 centrifugal filters, EMD Millipore, Burlington, MA. USA) for the first fraction, then centrifuged at 3234× *g* until the volume of retentate reached 250 µL. After collecting the retentate, the <30 kD filtrate was transferred to a 10 kD MWCO filter device for the second fraction. After centrifugation at 3234× *g*, 250 µL of retentate was collected. The <10 kD filtrate was added to the 3 kD MWCO filter device for the third fractionation and was centrifuged at 3234× *g* until the retentate volume reached 250 µL. Both the retentate and the <3 kD fraction were collected and all the hydrolysate fractions were stored at −80 °C until used for further assays.

2.8. Rat Intestine α-Glucosidase Inhibition Assay

Rat intestine α-glucosidase inhibition assay was conducted according to the protocol in the Worthington Enzyme Manual with modifications [32] and Kwon et al. [33]. Crude enzyme was extracted from rat intestine acetone powder. For the extraction, 0.3 g of rat intestinal acetone powder was added to 12 mL of 0.1 M sodium phosphate buffer (pH 6.9 with 0.9% NaCl), then sonicated 12 times in 30 s pulses. After centrifugation at 10,000× *g* for 30 min at 4 °C, the supernatant was used as the enzyme solution. A volume of 50 µL of solubilized sample or acarbose (positive control) and 100 µL of enzyme solution were added in a 96 well plate then incubated at 37 °C for 10 min. Then, 50 µL of 5 mM p-NPG solution in 0.1 M phosphate buffer (pH 6.9 with 0.9% NaCl) was added and the mixture was incubated at 37 °C for 30 min. The absorbance was measured at 405 nm by a microplate reader (Ex 808, Biotek, Winooski, VT, USA) and compared with a control containing 50 µL of 0.1 M sodium phosphate buffer in place of the sample. The α-glucosidase inhibitory activity was calculated as follows:

$$\% \text{ inhibition} = \frac{\left(\Delta A_{405}^{\text{Control}} - \Delta A_{405}^{\text{sample}}\right)}{\Delta A_{405}^{\text{Control}}} \times 100 \quad (4)$$

2.9. Porcine Pancreatic α-Amylase Inhibition Assay

The α-amylase inhibitory activity was evaluated following a modified version of the assay described by the Worthington Enzyme Manual [34] and Kwon et al. [26]. Porcine pancreatic α-amylase solution (100 unit/mL) was prepared in 0.02 M sodium phosphate buffer (pH 6.9 with 6 mM NaCl). A volume of 100 μL of solubilized sample or acarbose (positive control) and 100 μL of enzyme solution was incubated at 25 °C for 10 min, then 100 μL of 1% starch solution in 0.02 M sodium phosphate buffer (pH 6.9 with 6 mM NaCl) was added. After incubation at 25 °C for 10 min, 200 μL of dinitrosalicylic acid color reagent was added, and, subsequently, the mixture was boiled in a water bath for 5 min. After cooling, 50 μL of the mixture was diluted with 200 μL of deionized water and the absorbance was measured at 540 nm by a microplate reader (Ex 808, Biotek, Winooski, VT, USA). The measured absorbance was compared with a control containing 100 μL of 0.02 M sodium phosphate buffer in place of the sample and the α-amylase inhibitory activity was calculated as follows:

$$\% \text{ inhibition} = \frac{\left(\Delta A_{540}^{\text{Control}} - \Delta A_{540}^{\text{sample}}\right)}{\Delta A_{540}^{\text{Control}}} \times 100 \quad (5)$$

2.10. DPP-IV Inhibition Assay

The DPP-IV inhibitory activity was determined using a method from Li-Chan et al. [35]. DPP-IV enzyme was purchased from EMD Millipore (Burlington, MA, USA) and 0.01 unit/mL of enzyme solution was prepared in 100 mM Tris-HCl buffer (pH 8.0). A total of 25 μL of sample and sitagliptin (positive control) and 25 μL of 1.59 mM Gly-Pro-p-nitroanilide solution in 100 mM Tris-HCl buffer (pH 8.0) was preincubated at 37 °C for 10 min, then 50 μL of enzyme solution was added. After incubation at 37 °C for 60 min, 100 μL of 1 M sodium acetate solution was added to stop the reaction. The absorbance was determined at 405 nm by a microplate reader (Ex 808, Biotek, Winooski, VT, USA). The absorbance was compared with a control of 25 μL of 100 mM Tris-HCl buffer (pH 8.0) instead of the sample solution. The DPP-IV inhibitory activity was expressed as percentage inhibition and was calculated as follows:

$$\% \text{ inhibition} = \frac{\left(\Delta A_{405}^{\text{Control}} - \Delta A_{405}^{\text{sample}}\right)}{\Delta A_{405}^{\text{Control}}} \times 100 \quad (6)$$

2.11. GLUTag Cell Culture

Murine GLUTag cells that are widely used for the stimulation of GLP-1 secretion were used in this study. GLUTag cells were gifted from Dr. D.J. Drucker (University of Toronto, Toronto, Canada) and cultured in Dulbecco's Modified Eagle Medium (DMEM, Invitrogen, Carlsbad, CA, USA) containing 5.5 mM glucose (pH 7.4) supplemented with 10% (*v*/*v*) fetal bovine serum, 100 U/mL penicillin-streptomycin, and 2 mM L-glutamine in a 75 cm^2 culture flask at 5% CO_2 and 37 °C and GLUTag cells at 80% confluency were regularly trypsinized for subculture.

2.12. Cell Viability Assay

Viability of the GLUTag cells was evaluated using the MTT (3-(4,5-dimethylthiazol-2-yl)-2,5-diphenyl tetrazolium bromide) assay. The surface of 96-well culture plates was coated with poly-D-lysine solution. A density of 2×10^4 cells was seeded in each well and incubated at 37 °C, 5% CO_2, and 80% humidity for 48 h. A volume of 20 μL sample solution (final concentration: 0.2–10 mg/mL) was added to the cells and incubated for 24 h. Then, the mixture was replaced with 100 μL of 0.5 mg/mL MTT reagent and incubated at 37 °C and 5% CO_2 for 4 h. The MTT reagent was removed, then the precipitate was solubilized in 100 μL of DMSO. The absorbance was measured at 540 nm by a microplate reader (Ex 808, Biotek, Winooski, VT, USA) and compared with a control containing 20 μL of ultrapure water in place of the sample solution. The cell viability was calculated as follows:

$$\% \text{ cell viability} = \frac{\left(\Delta A_{540}^{\text{Sample}} - \Delta A_{540}^{\text{blank}}\right)}{\left(\Delta A_{540}^{\text{Control}} - \Delta A_{540}^{\text{blank}}\right)} \times 100 \quad (7)$$

2.13. GLP-1 Secretion Assay

GLP-1 secretory activity was evaluated by an assay modified from Ojo et al. [36]. GLUTag cells were seeded at a density of 2×10^5 in 24-well culture plates coated with poly-D-lysine and grown for 48 h in a 37 °C and 5% CO_2 incubator until the confluency reached 80%. The cells were washed with phosphate-buffered saline (PBS), then 1 mL of Kreb Ringer Buffer (KRB) containing 115 mM NaCl, 4.7 mM KCl, 1.28 mM $CaCl_2$, 1.2 mM $MgSO_4$, 1.2 mM KH_2PO_4, 25 mM HEPES, and 1M $NaHCO_3$ supplemented with 1% bovine serum albumin (BSA, fatty-acid-free) and 1 mM glucose (pH 7.4) was added. After pre-incubation at 37 °C and 5% CO_2 for 45 min, the KRB was replaced with 1 mL of appropriate KRB test solutions containing sample or 10 mM L-glutamine (positive control). All the KRB test solutions were prepared with supplementation of 1% BSA and 2 mM glucose (pH 7.4). The culture plates were incubated at 37 °C and 5% CO_2 for 2 h, then the mixture was collected. After centrifugation at 4 °C and 800× *g* for 5 min, the supernatant was collected and stored at −80 °C until further evaluation of GLP-1 concentration. Total GLP-1 concentration was determined based on the manufacturer's instructions in a commercial ELISA kit (GLP-1 Total ELISA, Millipore, Burlington, MA, USA). The GLP-1 secretory activity was expressed as a percentage (%) of the negative control (KRB buffer).

2.14. Statistical Analysis

The enzymatic hydrolysis process using each of the four commercial proteases was replicated three times and all of the assays were conducted in triplicate on each sample replicate. Statistical differences among the means of each treatment were evaluated using one-way analysis of variance (ANOVA) followed by Tukey's HSD post hoc test and paired *t*-test with a significance value of $p < 0.05$ (SPSS ver. 23, IBM Corp., Armonk, NY, USA). Correlations ($p < 0.05$) between the degree of hydrolysis (DH) and each biofunctional activity were analyzed through the Pearson coefficient (SPSS ver. 23, IBM Corp., Armonk, NY, USA).

3. Results

3.1. Degree of Hydrolysis

The degree of hydrolysis (DH) was determined by the OPA method with serine as a standard. DH was significantly increased after enzymatic hydrolysis for 60 min. DH of green crab hydrolysates ranged from 15.8% to 18.4%, while DH of the CMC (crab mince control) was 6.8%. Among the treatments, AL showed the highest DH (18.4% ± 0.4%) followed by the PR (17.1% ± 0.2%), FL (16.5% ± 1.2%), and PA (15.8% ± 0.3%) treatments, respectively (Figure 2).

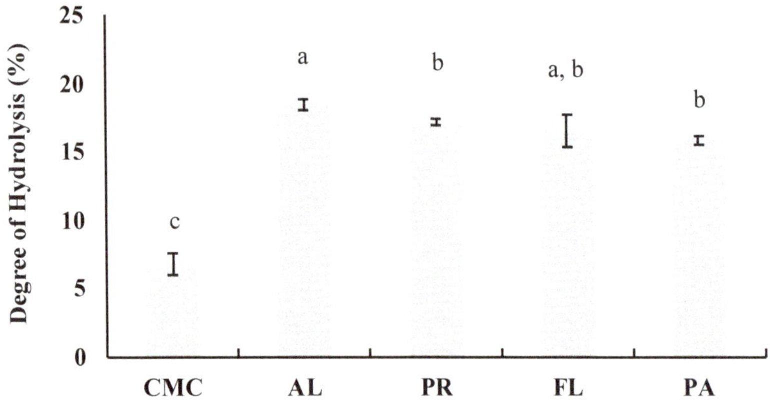

Figure 2. Degree of hydrolysis of green crab mince by enzyme type. CMC: Crab Mince Control, AL: Alcalase, PR: Protamex, FL: Flavourzyme, and PA: Papain. Each bar indicates the mean and standard deviation. Different letters on the bars indicate a significant difference ($p < 0.05$) among the treatments ($n = 3$).

3.2. Molecular Weight Distribution Using SDS-PAGE

The size distribution of proteins and peptides in the CMC and hydrolysates was determined using SDS-PAGE (Figure 3a). Significant hydrolysis was observed in all of the treatments compared to the CMC. Heavy chains in the 150–250 kD range were reduced by all the enzyme treatments. There were thick bands between 100 kD and 75 kD in the hydrolysates derived by FL and PA, and all of the treatments produced various bands between 20 kD and 37 kD. Similar band patterns were observed in the AL and PR treatments while FL and PA treatments resulted in similar molecular weight profiles. AL-derived hydrolysate showed the lowest intensity of bands between 2 and 10 kD compared to the CMC and other hydrolysates. During simulated digestion, the CMC was hydrolyzed by digestive enzymes as compared to the process control that was applied to the simulated digestion with no digestive enzymes (Figure 3b). The simulated digestion process further hydrolyzed all the hydrolysates obtained from commercial enzyme treatments. The intensity of bands above 15 kD decreased significantly as compared to before the simulated digestion, and the intensity of observed bands at <10 kD also decreased after the simulated digestion.

3.3. α-Glucosidase and Porcine α-Amylase Inhibitory Activities

The inhibitory activity of rat intestinal α-glucosidase and porcine pancreatic α-glucosidase was assessed to evaluate the potential anti-hyperglycemic effect of the CMC and crab hydrolysates. IC_{50} values of both α-glucosidase and α-amylase were determined to evaluate the efficacy of the inhibition. Acarbose was used as a positive control. Acarbose had IC_{50} values of 0.027 mg/mL and 0.84 mg/mL for α-glucosidase and α-glucosidase inhibitory activity, respectively. The α-glucosidase inhibitory activity of the CMC was significantly improved by the PR (IC_{50} 1.38 ± 0.19 mg/mL) and PA (IC_{50} 5.56 ± 0.19 mg/mL) treatments while the FL (IC_{50} 20.24 ± 0.19 mg/mL) treatment showed a decrease in α-glucosidase inhibitory activity compared to the CMC (IC_{50} 8.54 ± 0.50 mg/mL) (Figure 4a). Simulated digestion did not significantly affect the α-glucosidase inhibitory activity of hydrolysates obtained from the PR (IC_{50} 1.49 ± 0.08 mg/mL) and PA (IC_{50} 5.31 ± 0.51 mg/mL) treatments. However, the inhibitory activity of the FL-derived hydrolysate was significantly improved after the simulated digestion (IC_{50} 10.61 ± 1.90 mg/mL). The green crab hydrolysates obtained from PR showed the highest α-glucosidase inhibitory activity before and after simulated digestion and the activity was approximately 4–14 times higher compared to other enzyme treatments and CMC based on IC_{50} values.

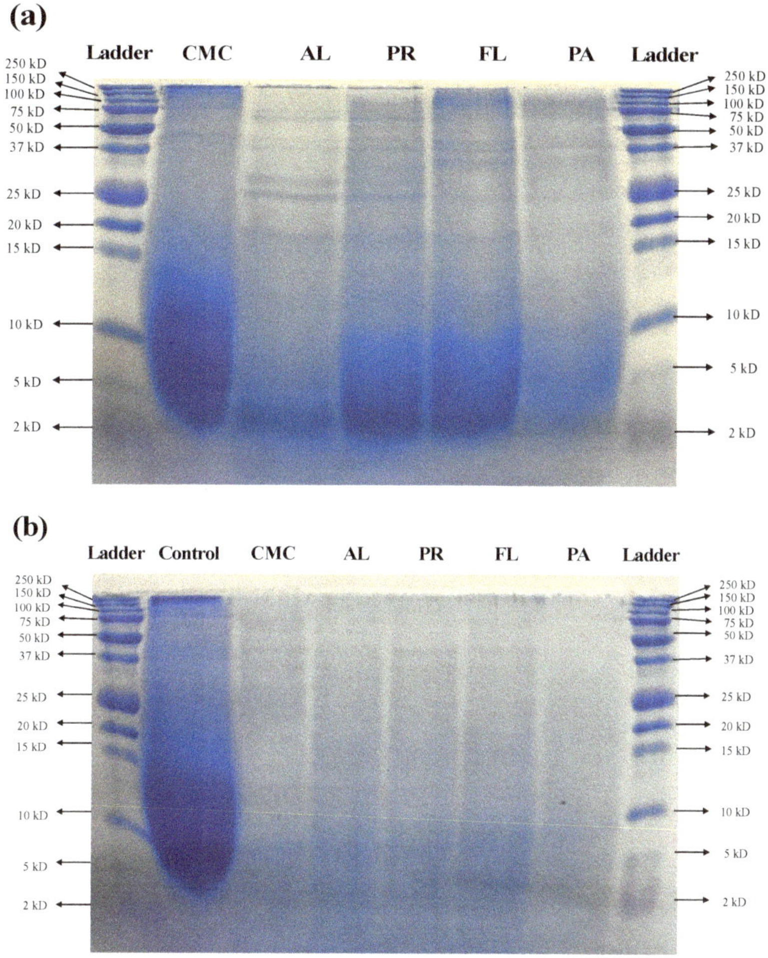

Figure 3. Molecular weight distribution (SDS-PAGE) of CMC and hydrolysates before and after simulated digestion. (**a**) Before simulated digestion. Lane 1 and 7: molecular weight ladder. Lanes 2–6: CMC, AL-derived hydrolysate, PR-derived hydrolysate, FL-derived hydrolysate, and PA-derived hydrolysate, respectively. (**b**) After simulated digestion. Lanes 1 and 8: molecular weight ladder. Lanes 2–7: simulated digestion process control, CMC, AL-derived hydrolysate, PR-derived hydrolysate, FL-derived hydrolysate, and PA-derived hydrolysate, respectively. CMC: Crab Mince Control, AL: Alcalase, PR: Protamex, FL: Flavourzyme, and PA: Papain.

The porcine pancreatic α-amylase inhibitory activity of the CMC was statistically increased by enzymatic hydrolysis using commercial enzymes. Among the treatments, PA (IC_{50} 9.35 ± 0.42 mg/mL) exhibited the highest α-amylase inhibitory activity followed by PR (IC_{50} 11.02 ± 0.69 mg/mL), FL (IC_{50} 11.12 ± 0.37 mg/mL), AL (IC_{50} 12.53 ± 0.85 mg/mL), and the CMC (IC_{50} 16.49 ± 0.41 mg/mL), respectively (Figure 4b). Digestive enzymes including pepsin and pancreatin significantly improved the α-amylase inhibitory activity of CMC (IC_{50} 10.55 ± 0.68 mg/mL) and its hydrolysates treated with AL (IC_{50} 10.11 ± 0.34 mg/mL), PR (IC_{50} 9.93 ± 0.82 mg/mL), and FL (IC_{50} 9.31 ± 0.37 mg/mL), however, the inhibitory activity of PA-derived hydrolysate was not affected by simulated digestion.

After simulated digestion, the α-amylase inhibitory activity among the treatments was not significantly different.

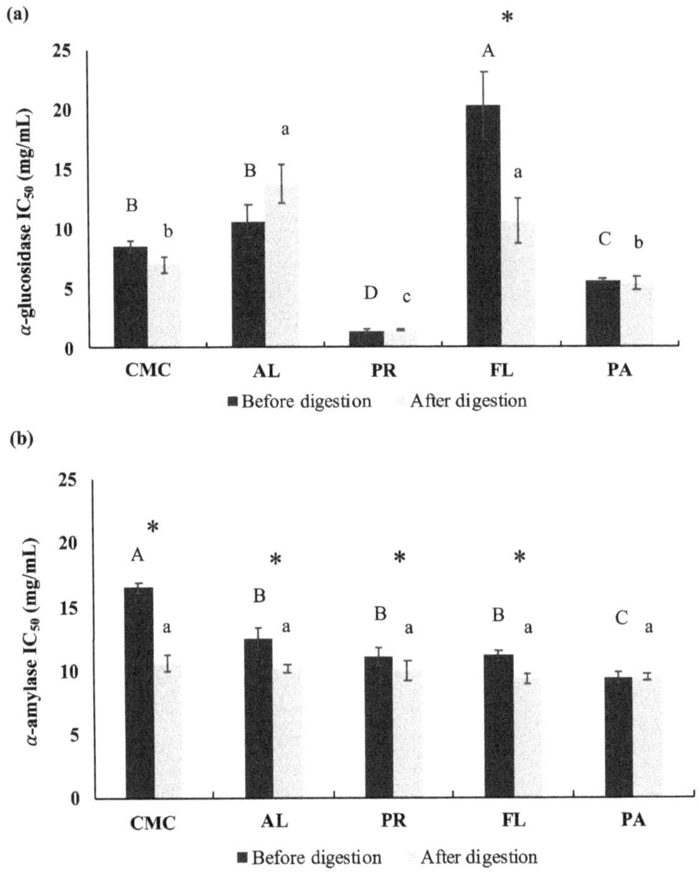

Figure 4. Rat intestinal α-glucosidase and porcine α-amylase inhibitory activities of CMC and hydrolysates before and after simulated digestion. (**a**) Rat intestinal α-glucosidase inhibitory activity. (**b**) Porcine α-amylase inhibitory activity. The IC_{50} values were represented by final assay concentration (protein and peptide basis). CMC: Crab Mince Control, AL: Alcalase, PR: Protamex, FL: Flavourzyme, and PA: Papain. Each bar indicates the mean and standard deviation (n = 3 treatment replicates). The letters on the bars indicate a significant difference ($p < 0.05$) among the treatments. An asterisk (*) represents significant difference after simulated digestion ($p < 0.05$) by paired t-test.

3.4. DPP-IV Inhibitory Activity

Overall, DPP-IV inhibition of CMC (IC50 1.50 ± 0.25 mg/mL) was significantly improved by the enzymatic hydrolysis, and the DPP-IV inhibitory activity IC50 values of hydrolysates ranged from 0.56 mg/mL to 0.72 mg/mL (Figure 5). Sitagliptin was used as a positive control, and had an IC_{50} value of 43.7 ng/mL. The simulated gastrointestinal digestion remarkably enhanced the DPP-IV inhibitory activity of CMC. However, the PR and Papain-treated samples showed significantly higher DPP-IV inhibitory activity than CMC after the simulated digestion.

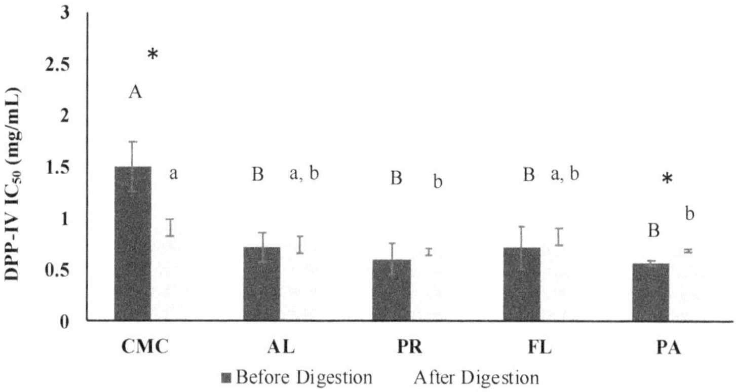

Figure 5. Dipeptidyl peptidase-IV (DPP-IV) IC_{50} values of CMC and hydrolysates. The IC_{50} values were represented by the final assay concentration (protein and peptide basis). CMC: Crab Mince Control, AL: Alcalase, PR: Protamex, FL: Flavourzyme, and PA: Papain. Each bar indicates the mean and standard deviation ($n = 3$ treatment replicates). The letters on the bars indicate a significant difference ($p < 0.05$) among the treatments. An asterisk (*) represents a significant difference after simulated digestion ($p < 0.05$) by paired *t*-test.

3.5. GLP-1 Secretory Activity

The GLP-1 secretory activity was measured at a sample concentration of 0.5 mg/mL. The cytotoxicity of the CMC and hydrolysates was determined using the MTT assay before evaluating their GLP-1 secretory activity. As a result, there was no observed cytotoxicity on GLUTag cells at 0.125, 0.25, and 0.5 mg/mL (Figure S1). The GLP-1 levels induced by the CMC and crab hydrolysate treatments were evaluated using the GLUTag cell model. Next, the results were compared with glutamine as a positive control (Figure 6). The GLP-1 secretion induced by the CMC and hydrolysates was significantly higher than that by the positive control by 2.5–3.5 times. Among the samples, AL-derived hydrolysate showed the lowest GLP-1 secretory activity, and the activity of all the samples was not significantly affected by simulated gastrointestinal digestion.

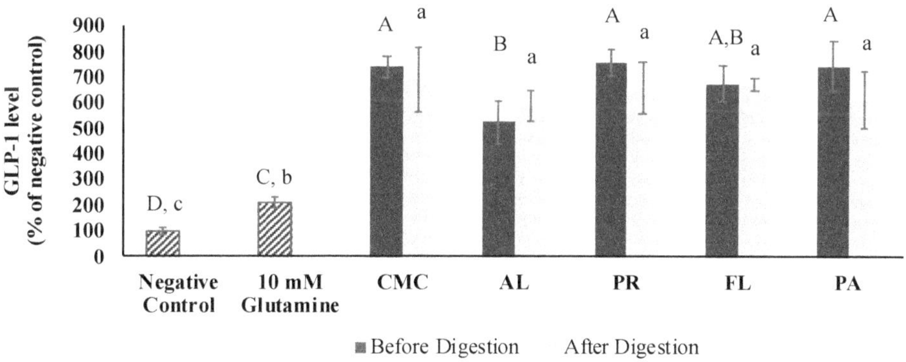

Figure 6. Glucagon-like 1 (GLP-1) level of CMC and hydrolysates before and after simulated digestion determined using GLUTag cells. The GLP-1 level was measured using 0.50 mg/mL of samples (final assay concentration, protein, and peptide basis). CMC: Crab Mince Control, AL: Alcalase, PR: Protamex, FL: Flavourzyme, and PA: Papain. Each bar indicates the mean and standard deviation ($n = 3$ treatment replicates). The letters on the bars indicate a significant difference ($p < 0.05$) among the treatments.

3.6. α-Glucosidase Inhibitory Activity of Fractions

Based on the positive results of their α-glucosidase and GLP-1 secretory activities, PR, FL, and PA were selected for subsequent fractionation. Figure 7 represents the α-glucosidase inhibitory activity of the fractions determined at 3.2 mg/mL. PR-treated fractions exhibited the highest α-glucosidase inhibitory activity among the treatments. Within the fractions, the highest α-glucosidase inhibitory activity was shown in the <3 kD fractions of all of the treatments. The <3 kD fraction of the PR-derived hydrolysate inhibited the α-glucosidase activity by 60% while the α-glucosidase inhibitory activity of other fractions was less than 50%. IC_{50} values could only be calculated for the <3 kD fraction since the highest inhibitory activity of the >3 kD fractions was less than 50%. The PR treatment (IC_{50} 1.75 ± 0.24 mg/mL) resulted in the highest α-glucosidase inhibitory effect followed by PA, CMC, and FL.

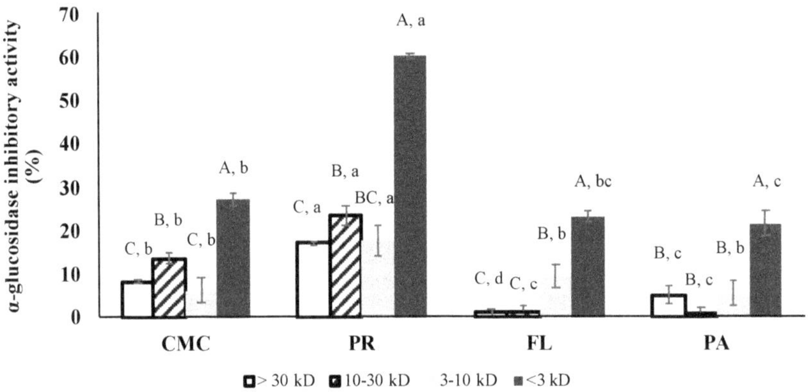

Figure 7. Rat intestinal α-glucosidase inhibitory activities of CMC and hydrolysates after fractionation (final assay concentration: 3.2 mg/mL protein and peptide basis). CMC: Crab Mince Control, AL: Alcalase, PR: Protamex, FL: Flavourzyme, and PA: Papain. Each bar indicates the mean and standard deviation (n = 3 treatment replicates). Uppercase letters on the bars indicate a significant difference ($p < 0.05$) within each treatment. Lowercase letters on the bars indicate a significant difference ($p < 0.05$) among the treatments.

3.7. α-Amylase Inhibitory Activity of Fractions

The α-amylase inhibition of all the fractions was evaluated at 4.2 mg/mL (Figure 8). The <3 kD fractions had the highest inhibitory activity among the fractions. The inhibitory activities of the <3 kD fractions were not significantly different between PR and FL treatments, which were significantly higher than the α-amylase inhibitory activities of the <3 kD fractions of CMC and PA-derived hydrolysates. IC_{50} values of all the <3 kD fractions ranged from 8.36 to 10.88 mg/mL and were not significantly different among the treatments.

3.8. DPP-IV Inhibitory Activity of Fractions

The DPP-IV inhibitory ability of the fractions obtained from the fractionation of the ±CMC and hydrolysates is shown in Figure 9. According to IC_{50} values, the <3 kD fraction (CMC: 1.39 ± 0.16 mg/mL, PR: 0.79 ± 0.05 mg/mL, FL: 0.88 ± 0.17 mg/mL, and PA: 0.93 ± 0.20 mg/mL) and 3–10 kD fraction (CMC: 1.68 ± 0.28 mg/mL, PR: 0.95 ± 0.06 mg/mL, FL: 1.15 ± 0.16 mg/mL, and PA:0.88 ± 0.17 mg/mL) within each sample showed the highest DPP-IV inhibitory activity while the > 30 kD fraction (CMC: 4.80 ± 0.37 mg/mL, PR: 2.29 ± 0.40 mg/mL, FL: 2.70 ± 0.26 mg/mL, and PA: 2.21 ± 0.20 mg/mL) exhibited the lowest DPP-IV inhibitory activity within enzyme treatments. The DPP-IV inhibitory activity of all the fractions was improved after the commercial enzyme treatments; however, there was no significant difference among the treatments.

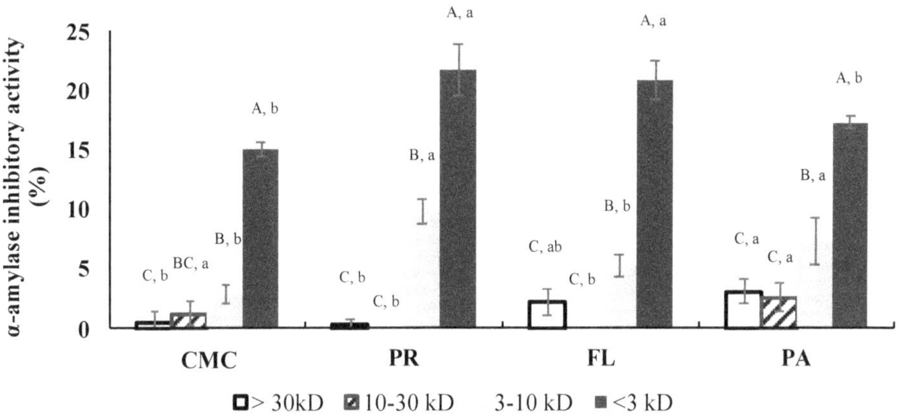

Figure 8. Porcine pancreatic α-amylase inhibitory activities of CMC and hydrolysates after fractionation (final assay concentration: 4.2 mg/mL protein and peptide basis). CMC: Crab Mince Control, AL: Alcalase, PR: Protamex, FL: Flavourzyme, and PA: Papain. Each bar indicates the mean and standard deviation (n = 3 treatment replicates). Uppercase letters on the bars indicate a significant difference ($p < 0.05$) within each treatment. Lowercase letters on the bars indicate a significant difference ($p < 0.05$) among the treatments.

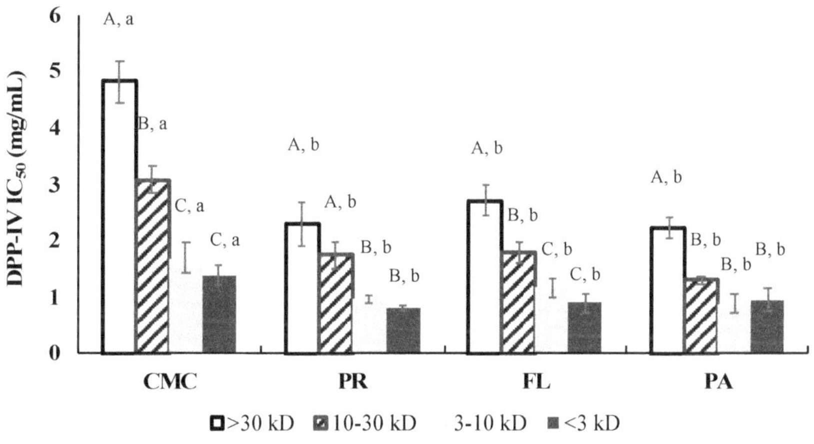

Figure 9. DPP-IV IC_{50} values of CMC and hydrolysates after fractionation. The IC_{50} values were represented by final assay concentration (protein and peptide basis). CMC: Crab Mince Control, AL: Alcalase, PR: Protamex, FL: Flavourzyme, and PA: Papain. Each bar indicates the mean and standard deviation (n = 3 treatment replicates). Uppercase letters on the bars indicate a significant difference ($p < 0.05$) within each treatment. Lowercase letters on the bars indicate a significant difference ($p < 0.05$) among the treatments.

3.9. GLP-1 Secretory Activity of Fractions

After the fractionation, the total GLP-1 level induced by the CMC and crab hydrolysates (0.5 mg/mL) was evaluated using GLUTag cells. The GLP-1 secretory activity of fractions was improved by the enzymatic hydrolysis except for the >30 kD fraction of PA-derived hydrolysate and <3 kD fractions of

FL and PA-derived hydrolysates (Figure 10). Glutamine (10 mM, positive control) increased GLP-1 level by 216% ± 13% compared to a negative control (KRB buffer), and the <3 kD fractions had significantly higher GLP-1 levels than the positive control.

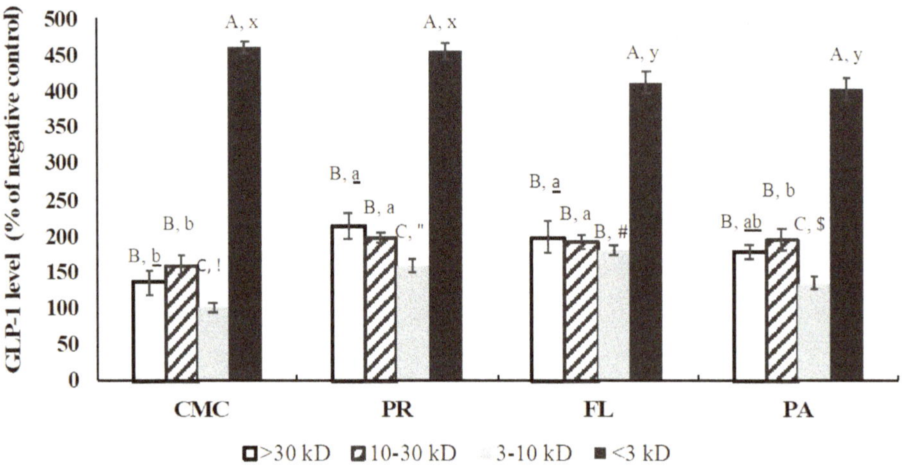

Figure 10. GLP-1 level of CMC and hydrolysates after fractionation (final assay concentration: 0.5 mg/mL protein and peptide basis). CMC: Crab Mince Control, AL: Alcalase, PR: Protamex, FL: Flavourzyme, and PA: Papain. Each bar indicates the mean and standard deviation (n = 3 treatment replicates). Lowercase, lowercase letters with underline, and Greek letters on the bars indicate a significant difference ($p < 0.05$) within each treatment. Uppercase letters on the bars indicate a significant difference ($p < 0.05$) among the treatments.

4. Discussion

In recent years, as interest in healthy lifestyles has increased, the demand for healthy diets, natural food ingredients, and bioactive compounds has intensified in consumers as well as the food industry. Proteins and peptides are being widely used as nutraceutical food ingredients in food products and dietary supplements to enhance human health and product value. Many bioactive peptides having antioxidant, anti-hypertensive, and anti-hyperglycemic effects have been derived from enzymatic hydrolysis of protein sources including meat, dairy, fish, and their byproducts [15–20,37–39]. However, there has been limited investigation into bioactive peptides derived from molluscan and crustacean resources.

The bioactive properties of peptides obtained by enzymatic hydrolysis are influenced by (1) the primary amino acid sequence of the protein source, (2) the specificity of the proteases applied, and (3) process parameters such as DH, hydrolysis duration, and enzyme:substrate ratio. As a primary protein source, underutilized green crab was used, which contains ~80% protein in its meat [7]. Manual removal of green crab meat from the carapace is extremely labor-intensive because of the crab's very small size. Therefore, in this study, a mechanical separator was used to generate green crab meat mince that was subsequently used as the substrate for the enzymatic hydrolysis. To apply proteases with various specificity to the crab homogenate, four commercially available proteases that have two different modes of action were selected. AL, PR, and PA are endopeptidases, while FL has both endo- and exopeptidase characteristics. After hydrolysis, the DH was determined based on the number of cleaved peptide bonds over the total number of peptide bonds. Based on our results, DH significantly increased in response to enzyme treatment (Figure 2). According to Jamdar et al. [40], the bioactivity of protein hydrolysates was modified as DH was changed within the same enzyme treatment. Therefore, the determination of DH is necessary for controlling the enzymatic hydrolysis process to reproducibly obtain bioactive peptides.

In addition to the increase in DH, the molecular weight distribution of the CMC and hydrolysates before the simulated digestion also confirms that the enzymatic process was successful, and that different products were generated based on the commercial protease applied (Figures 2 and 3). Although similar MW patterns were observed in AL and PR-treated samples, their activities were significantly different in terms of α-glucosidase inhibition and GLP-1 secretion (Figures 3 and 5). The weak intensity of AL-derived hydrolysate bands between 2 and 10 kD may be associated with its DH (Figure 3a). In many studies, AL showed an outstanding ability to hydrolyze large proteins to small peptides, and produced smaller peptides more rapidly compared to other commercially available proteases [41,42]. Therefore, the lower intensity of the 2–10 kD bands in the AL treatment was most likely due to a large number of peptides smaller than <2 kD passing through the gel during the SDS-PAGE. After the simulated digestion, the amount of larger size peptides was decreased, and the overall intensity of all the bands became weaker (Figure 3a,b). This demonstrates that proteins and peptides in the CMC and hydrolysates were further hydrolyzed by the digestive enzymes and suggests that the smaller sized peptides may have passed through the gel during electrophoresis.

Pancreatic α-amylase and intestinal α-glucosidase are the two key enzymes involved in starch digestion, resulting in an increase in blood glucose levels. Therefore, the inhibition of both carbohydrases can be an indicator of the potential anti-hyperglycemic effect [43]. According to the DH and SDS-PAGE results, AL hydrolyzed the green crab protein to a greater degree compared to the other enzymes (Figures 2 and 3a). However, results indicate that PR generated the most effective α-glucosidase inhibitive hydrolysates (Figure 4), and that the inhibitory activity was not statistically correlated ($p > 0.05$) with DH. The IC_{50} values of the control and the hydrolysates were strongly correlated with the <3 kD fractions ($r = 0.987$, $p < 0.05$), which indicates that the peptides in the <3 kD fraction are key contributors to the α-glucosidase inhibitory activity. The α-glucosidase inhibitory activity of PR-treated sample (IC_{50} 1.38 ± 0.19 mg/mL) was remarkably higher than not only other treatments and the control but also the hydrolysates of sardine muscle (IC_{50} 48.7 mg/mL), whey protein isolate (IC_{50} 4.5 mg/mL), and edible insects (IC_{50} > 2.0 mg/mL) including mealworm larvae, crickets, and silkworm pupae [38,44,45]. In many studies, AL, FL, and digestive enzymes including pepsin and trypsin are commonly used to obtain bioactive peptides [38,39,45,46]. Interestingly, in the current study, AL and FL treatment did not show improved inhibitory activity compared to the control. The high α-glucosidase inhibitory activity of the PR-treated sample emphasizes the importance of protein source and protease selection in obtaining bioactive peptides. A major concern regarding the use of bioactive peptides for human consumption and industrial application are the changes in biological activity as a result of further hydrolysis during gastrointestinal digestion and processing. However, simulated digestion did not affect the α-glucosidase inhibitory activities of PR and PA-treated samples, suggesting that their inhibitory activity was stable to pepsin and pancreatin action during the gastrointestinal digestion.

The enzymatic hydrolysis by commercial enzymes, and subsequently by digestive enzymes, improved the α-amylase inhibitory activity of the green crab proteins. The inhibitory activity was most likely primarily due to the <3 kD fractions (Figures 4b and 8) since those showed the highest inhibitory activity among the same concentration of fractions. The α-amylase inhibition of the PR-treated sample was weaker than its α-glucosidase inhibition which may contribute to reducing the side-effects that are frequently caused by inhibition of carbohydrate hydrolyzing enzymes. Acarbose, a synthetic pharmaceutical α-glucosidase and α-amylase inhibitor, has been commonly used to inhibit glucose absorption in diabetic patients. However, its strong α-amylase inhibitory activity (IC_{50} <1 mg/mL) leads to the presence of non-digested polysaccharides in the large intestine, which causes side-effects including severe stomach pain, constipation, and diarrhea [47]. Therefore, the use of α-glucosidase and α-amylase inhibitors based on the PR-derived hydrolysates might be a good alternative to delay glucose absorption and help control blood glucose spikes. The inhibitory mechanism of both α-glucosidase and α-amylase by bioactive peptides has not been well characterized. However, recent studies have reported that hydrophobic interactions of non-saccharide compounds that allow them to bind to the carbohydrase active site may contribute to their inhibitory activity [48].

DPP-IV is an enzyme that rapidly metabolizes incretins such as active GLP-1 and gastric inhibitory polypeptide (GIP) hormones. These incretins are important since the hormones help in blood glucose control by insulin secretion [49]. DPP-IV as a postproline hydrolyzing enzyme cleaves dipeptides with X-Pro or X-Ala from the N-terminus of polypeptides [50,51]. In this study, the DPP-IV inhibitory activity of the CMC and hydrolysates was investigated as a potential strategy for T2DM management. Enzyme hydrolysis improved the DPP-IV inhibitory activity regardless of the protease applied, and the activity was not significantly ($p > 0.05$) correlated with DH. Recent studies on DPP-IV inhibitors derived from food sources reported that peptides containing Pro, Ala, and Gly at the P1-position and/or Trp at the N-terminal might have effective DPP-IV inhibitory activity [52]. The type of amino acid residues at the P1, P2, and P1′-positions of peptides may significantly influence their DPP-IV inhibitory activity [53]. The <3 kD and 3–10 kD fractions showed similar IC_{50} values to the hydrolysates before the simulated digestion (Figures 5 and 9). Thus, our study suggests that the fractions <10 kD are likely to play an important role in inhibiting the DPP-IV enzyme. However, overall DPP-IV inhibitory activity of the CMC and hydrolysates (IC_{50} 0.56–1.5 mg/mL) was significantly lower than the medication sitagliptin (IC_{50} 43.7 ng/mL) and other dairy protein and salmon byproduct hydrolysates (IC_{50} < 100 μg/mL) [35,38].

GLP-1 hormone has the ability to reduce blood sugar levels by enhancing the secretion of insulin; however, it has a half-life of only ~1.5 min because of rapid inactivation by DPP-IV [54]. The GLP-1 content secreted from GLUTag cells was not different among the samples except for AL treatment that showed a lower GLP-1 level, and the GLP-1 secretory activity was not correlated with DH. In comparison with the positive control and hydrolysates from blue whiting and salmon skin, green crab hydrolysates released approximately 2.5–3.5 times higher GLP-1 levels. This is considered due to the mixture of certain free amino acids and peptides in the crab hydrolysates. Various amino acids including Gln, Glu, Ala, Ser, Leu, Gly, Asn, and Met have been demonstrated to stimulate GLP-1 secretory activity [39,55–58]. According to Tolhurst [59], there are two major mechanisms involved in the stimulation of GLP-1 release by Gln on Glutag cells: (1) electrogenic sodium-coupled amino acid uptake resulting in a depolarization of membrane and activation of voltage-gated calcium entry, and (2) elaboration of intracellular cAMP levels. Gameiro et al. [58]. reported that amino acids such as Ala and Gly activate ionotropic glycine receptors, which respond to the various amino acids by generating chloride current. Our previous research showed that the green crab mince contains a high amount of Gln + Glu and Asn + Asp [7], which may have contributed to its higher GLP-1 secretory activity compared to the negative and positive controls. Small peptides including tri- and dipeptides (Leu-Gly-Gly, Gly-Leu, and Gly- Pro) are known to stimulate the secretion of GLP-1 from the in vitro cell model [39,60]. Therefore, the high GLP-1 content induced by the <3 kD fractions within treatments might be due to the concentration of amino acids and small di- or tripeptides after the fractionation.

5. Conclusions

Enzymatic hydrolysis was successfully applied to mechanically separated green crab mince to generate a potential anti-hyperglycemic food ingredient. Our results suggest that the anti-hyperglycemic effects of green crab hydrolysates were dependent on the type of protease applied, and not on the degree of hydrolysis. Among the proteases evaluated, Protamex generated products having the highest α-glucosidase inhibitory activity. Furthermore, the fractionation study indicated that the <3 kD peptides primarily contributed to the bioactivities of the hydrolysates. Importantly for their prospective use in novel food products, the bioactivities of the hydrolysates were stable after the simulated gastrointestinal digestion. In conclusion, Protamex was the most effective protease for obtaining anti-hyperglycemic hydrolysates from green crab mince, and the development of bioactive hydrolysates may be a viable route for developing value-added food ingredients from this underutilized marine resource. However, substantial further research including identifying the bioactive compounds in the hydrolysates, evaluating their stability in a food model, and conducting in vivo assessment of their effectiveness is required for their future commercial application in health-promoting foods.

Supplementary Materials: The following are available online at http://www.mdpi.com/2304-8158/9/3/258/s1, Figure S1: Effect of CMC and hydrolysates on GLUTag cells. Each bar indicates the mean and standard deviation ($n = 3$ treatment replicates). Asterisk (*) represents significant difference between negative control and each sample ($p < 0.05$) by paired *t*-test.

Author Contributions: Conceptualization, B.K., D.I.S., and A.D.M.; methodology, B.K., D.I.S., and A.D.M.; validation, B.K., D.I.S., and A.D.M.; formal analysis, B.K.; investigation, B.K.; data curation, B.K.; writing—original draft preparation, B.K.; writing—review and editing, D.I.S., and A.D.M.; supervision, D.I.S., and A.D.M.; funding acquisition, D.I.S., and A.D.M. All authors have read and agreed to the published version of the manuscript.

Funding: This research was funded by National Science Foundation award, grant number 1355457 to Maine EPSCOR at the University of Maine.

Acknowledgments: We are thankful to Marissa McMahan at Manomet (Brunswick, ME) for providing green crabs. This manuscript is #3730 of the University of Maine Agricultural and Forest Experiment Station.

Conflicts of Interest: The authors declare no conflicts of interest.

References

1. Suleria, H.A.R.; Osborne, S.; Masci, P.; Gobe, G. Marine-based nutraceuticals: An innovative trend in the food and supplement industries. *Mar. Drugs* **2015**, *13*, 6336–6351. [CrossRef]
2. Kim, S.K.; Wijesekara, I. Development and biological activities of marine-derived bioactive peptides: A review. *J. Funct. Foods* **2010**, *2*, 1–9. [CrossRef]
3. Leignel, V.; Stillman, J.H.; Baringou, S.; Thabet, R.; Metais, I. Overview on the European green crab *Carcinus spp.* (*Portunidae, Decapoda*), one of the most famous marine invaders and ecotoxicological models. *Env. Sci. Pollut. Res.* **2014**, *21*, 9129–9144. [CrossRef]
4. Sigurdsson, G.M.; Rochette, R. Predation by green crab and sand shrimp on settling and recently settled American lobster postlarvae. *J. Crustac. Biol.* **2013**, *33*, 10–14. [CrossRef]
5. Malyshev, A.; Quijón, P.A. Disruption of essential habitat by a coastal invader: New evidence of the effects of green crabs on eelgrass beds. *Ices. J. Mar. Sci.* **2011**, *68*, 1852–1856. [CrossRef]
6. Grosholz, E.D.; Ruiz, G.M. Predicting the impact of introduced marine species: Lessons from the multiple invasions of the European green crab *Carcinus maenas*. *Biol. Conserv.* **1996**, *78*, 59–66. [CrossRef]
7. Kang, B.; Myracle, A.D.; Skonberg, D.I. Potential of recovered proteins from invasive green crabs (*Carcinus maenas*) as a functional food ingredient. *J. Sci. Food Agric* **2019**, *99*, 1748–1754. [CrossRef]
8. Galetti, J.A.; Calder, B.L.; Skonberg, D.I. Mechanical separation of green crab (*carcinus maenas*) meat and consumer acceptability of a value-added food product. *J. Aquat. Food Prod. T.* **2017**, *26*, 172–180. [CrossRef]
9. Naczk, M.; Williams, J.; Brennan, K.; Liyanapathirana, C.; Shahidi, F. Compositional characteristics of green crab (*Carcinus maenas*). *Food Chem.* **2004**, *88*, 429–434. [CrossRef]
10. Skonberg, D.I.; Perkins, B.L. Nutrient composition of green crab (*Carcinus maenus*) leg meat and claw meat. *Food Chem.* **2002**, *77*, 401–404. [CrossRef]
11. Baharuddin, N.A.; Halim, N.R.A.; Sarbon, N.M. Effect of degree of hydrolysis (DH) on the functional properties and angiotensin I-converting enzyme (ACE) inhibitory activity of eel (Monopterus sp.) protein hydrolysate. *Int. Food Res. J.* **2016**, *23*, 1424–1431.
12. Intarasirisawat, R.; Benjakul, S.; Visessanguan, W.; Wu, J. Antioxidative and functional properties of protein hydrolysate from defatted skipjack (*Katsuwonous pelamis*) roe. *Food Chem.* **2012**, *135*, 3039–3048. [CrossRef]
13. Sánchez, A.; Vázquez, A. Bioactive peptides: A review. *Food Qual. Saf.* **2017**, *1*, 29–46. [CrossRef]
14. Ndiaye, F.; Vuong, T.; Duarte, J.; Aluko, R.E.; Matar, C. Anti-oxidant, anti-inflammatory and immunomodulating properties of an enzymatic protein hydrolysate from yellow field pea seeds. *Eur. J. Nutr.* **2012**, *51*, 29–37. [CrossRef]
15. Pan, D.; Cao, J.; Guo, H.; Zhao, B. Studies on purification and the molecular mechanism of a novel ACE inhibitory peptide from whey protein hydrolysate. *Food Chem.* **2012**, *130*, 121–126. [CrossRef]
16. Slizyte, R.; Rommi, K.; Mozuraityte, R.; Eck, P.; Five, K.; Rustad, T. Bioactivities of fish protein hydrolysates from defatted salmon backbones. *Biotechnol. Rep.* **2016**, *11*, 99–109. [CrossRef]

17. Ahn, C.B.; Je, J.Y.; Cho, Y.S. Antioxidant and anti-inflammatory peptide fraction from salmon byproduct protein hydrolysates by peptic hydrolysis. *Food Res. Int.* **2012**, *49*, 92–98. [CrossRef]
18. Ahn, C.B.; Jeon, Y.J.; Kim, Y.T.; Je, J.Y. Angiotensin I converting enzyme (ACE) inhibitory peptides from salmon byproduct protein hydrolysate by Alcalase hydrolysis. *Process Biochem.* **2012**, *47*, 2240–2245. [CrossRef]
19. Ovissipour, M.; Rasco, B.; Shiroodi, S.G.; Modanlow, M.; Gholami, S.; Nemati, M. Antioxidant activity of protein hydrolysates from whole anchovy sprat (*Clupeonella engrauliformis*) prepared using endogenous enzymes and commercial proteases. *J. Sci. Food Agric.* **2013**, *93*, 1718–1726. [CrossRef]
20. Liu, F.R.; Wang, L.; Wang, R.; Chen, Z.X. Calcium-binding capacity of wheat germ protein hydrolysate and characterization of peptide–calcium complex. *J. Agr. Food Chem.* **2013**, *61*, 7537–7544. [CrossRef]
21. World Health Organization. The Top 10 Causes of Death. Available online: https://www.who.int/news-room/fact-sheets/detail/the-top-10-causes-of-death (accessed on 10 January 2020).
22. Centers for Disease Control and Prevention. Available online: https://www.cdc.gov/chronicdisease/about/costs/index.htm (accessed on 10 January 2020).
23. Ley, S.H.; Hamdy, O.; Mohan, V.; Hu, F.B. Prevention and management of type 2 diabetes: Dietary components and nutritional strategies. *Lancet* **2014**, *383*, 1999–2007. [CrossRef]
24. Steyn, N.P.; Mann, J.; Bennett, P.H.; Temple, N.; Zimmet, P.; Tuomilehto, J.; Louheranta, A. Diet, nutrition and the prevention of type 2 diabetes. *Public Health Nut.* **2004**, *7*, 147–165. [CrossRef]
25. Diabetes Prevention Program Research Group. Costs associated with the primary prevention of type 2 diabetes mellitus in the diabetes prevention program. *Diabetes Care* **2003**, *26*, 36–47. [CrossRef] [PubMed]
26. de Campos Zani, S.C.; Wu, J.; Chan, C.B. Egg and soy-derived peptides and hydrolysates: A review of their physiological actions against diabetes and obesity. *Nutrients* **2018**, *10*, 549. [CrossRef] [PubMed]
27. Beaulieu, L.; Thibodeau, J.; Bryl, P.; Carbonneau, M.É. Characterization of enzymatic hydrolyzed snow crab (Chionoecetes opilio) by-product fractions: A source of high-valued biomolecules. *Bioresour. Technol.* **2009**, *100*, 3332–3342. [CrossRef]
28. Nielsen, P.M.; Petersen, D.; Dambmann, C. Improved method for determining food protein degree of hydrolysis. *J. Food Sci.* **2001**, *66*, 642–646. [CrossRef]
29. Karayannakidis, P.D.; Apostolidis, E.; Lee, C.M. Comparison of direct steam injection and steam-jacketed heating in squid protein hydrolysis for energy consumption and hydrolysis performance. *Lwt Food Sci. Technol.* **2014**, *57*, 134–140. [CrossRef]
30. González-Montoya, M.; Hernández-Ledesma, B.; Silván, J.M.; Mora-Escobedo, R.; Martínez-Villaluenga, C. Peptides derived from in vitro gastrointestinal digestion of germinated soybean proteins inhibit human colon cancer cells proliferation and inflammation. *Food Chem.* **2018**, *242*, 75–82. [CrossRef]
31. Laemmli, U.K. Cleavage of structural proteins during the assembly of the head of bacteriophage T4. *Nature* **1970**, *227*, 680. [CrossRef]
32. Worthington, V. Maltase-α-glucosidase. In *Worthington Enzyme Manual*, 1st ed.; Worthington Biochemical Corp: Freehold, NJ, USA, 1993; p. 261.
33. Kwon, Y.I.; Apostolidis, E.; Kim, Y.C.; Shetty, K. Health benefits of traditional corn, beans, and pumpkin: In vitro studies for hyperglycemia and hypertension management. *J. Med. Food.* **2007**, *10*, 266–275. [CrossRef]
34. Worthington, V. α-amylase. In *Worthington Enzyme Manual*, 1st ed.; Worthington Biochemical Corp: Freehold, NJ, USA, 1993; pp. 36–41.
35. Li-Chan, E.C.; Hunag, S.L.; Jao, C.L.; Ho, K.P.; Hsu, K.C. Peptides derived from Atlantic salmon skin gelatin as dipeptidyl-peptidase IV inhibitors. *J. Agric. Food. Chem.* **2012**, *60*, 973–978. [CrossRef]
36. Ojo, O.O.; Conlon, J.M.; Flatt, P.R.; Abdel-Wahab, Y.H.A. Frog skin peptides (tigerinin-1R, magainin-AM1,-AM2, CPF-AM1, and PGla-AM1) stimulate secretion of glucagon-like peptide 1 (GLP-1) by GLUTag cells. *Biochem. Biophys. Res. Commun.* **2013**, *431*, 14–18. [CrossRef]
37. Chakka, A.K.; Elias, M.; Jini, R.; Sakhare, P.Z.; Bhaskar, N. In-vitro antioxidant and antibacterial properties of fermentatively and enzymatically prepared chicken liver protein hydrolysates. *J. Food Sci. Technol.* **2015**, *52*, 8059–8067. [CrossRef]
38. Lacroix, I.M.; Li-Chan, E.C. Inhibition of dipeptidyl peptidase (DPP)-IV and α-glucosidase activities by pepsin-treated whey proteins. *J. Agric. Food. Chem.* **2013**, *61*, 7500–7506. [CrossRef] [PubMed]
39. Harnedy, P.A.; Parthsarathy, V.; McLaughlin, C.M.; O'Keeffe, M.B.; Allsopp, P.J.; McSorley, E.M.; Allsopp, P.J.; McSorley, E.M.; O'Harte, F.P.M.; FitzGerald, R.J. Blue whiting (*Micromesistius poutassou*) muscle protein hydrolysate with in vitro and in vivo antidiabetic properties. *J. Funct. Foods.* **2018**, *40*, 137–145. [CrossRef]

40. Jamdar, S.N.; Rajalakshmi, V.; Pednekar, M.D.; Juan, F.; Yardi, V.; Sharma, A. Influence of degree of hydrolysis on functional properties, antioxidant activity and ACE inhibitory activity of peanut protein hydrolysate. *Food Chem.* **2010**, *121*, 178–184. [CrossRef]
41. Smyth, M.; FitzGerald, R.J. Relationship between some characteristics of WPC hydrolysates and the enzyme complement in commercially available proteinase preparations. *Int. Dairy J.* **1998**, *8*, 819–827. [CrossRef]
42. Aspmo, S.I.; Horn, S.J.; Eijsink, V.G. Enzymatic hydrolysis of Atlantic cod (*Gadus morhua L.*) viscera. *Process Biochem.* **2005**, *40*, 1957–1966. [CrossRef]
43. Hirsh, A.J.; Yao, S.Y.; Young, J.D. Inhibition of glucose absorption in the rat jejunum: A novel action of alpha-D-glucosidase inhibitors. *Gastroenterology* **1997**, *113*, 205–211. [CrossRef]
44. Matsui, T.; Oki, T.; Osajima, Y. Isolation and identification of peptidic α -glucosidase inhibitors derived from sardine muscle hydrolyzate. *Z. Nat. C* **1999**, *54*, 259–263. [CrossRef]
45. Yoon, S.; Wong, N.A.; Chae, M.; Auh, J.H. Comparative characterization of protein hydrolysates from three edible insects: Mealworm larvae, adult crickets, and silkworm pupae. *Foods* **2019**, *8*, 563. [CrossRef] [PubMed]
46. Ambigaipalan, P.; Al-Khalifa, A.S.; Shahidi, F. Antioxidant and angiotensin I converting enzyme (ACE) inhibitory activities of date seed protein hydrolysates prepared using Alcalase, Flavourzyme and Thermolysin. *J. Funct. Foods* **2015**, *18*, 1125–1137. [CrossRef]
47. Chiasson, J.L.; Josse, R.G.; Gomis, R.; Hanefeld, M.; Karasik, A.; Laakso, M. STOP-NIDDM Trial Research Group. Acarbose for prevention of type 2 diabetes mellitus: The STOP-NIDDM randomised trial. *Lancet* **2002**, *359*, 2072–2077. [CrossRef]
48. Bharatham, K.; Bharatham, N.; Park, K.H.; Lee, K.W. Binding mode analyses and pharmacophore model development for sulfonamide chalcone derivatives, a new class of α-glucosidase inhibitors. *J. Mol. Graph. Model.* **2008**, *26*, 1202–1212. [CrossRef] [PubMed]
49. Addison, D.; Aguilar, D. Diabetes and cardiovascular disease: Thepotential benefit of incretin-based therapies. *Curr. Atheroscler. Rep.* **2011**, *13*, 115–122. [CrossRef] [PubMed]
50. Blanco, J.; Nguyen, C.; Callebaut, C.; Jacotot, E.; Krust, B.; Mazaleyrat, J.P.; Wakselman, M.; Hovanessian, A.G. Dipeptidyl-peptidase IV-b. Further characterization and comparison to dipeptidyl- peptidase IV activity of CD26. *Eur. J. Biochem.* **1998**, *256*, 369–378. [CrossRef]
51. Hildebrandt, M.; Reutter, W.; Arck, P.; Rose, M.; Klapp, B.F. A guardian angel: The involvement of dipeptidyl peptidase IV in psychoneuroendocrine function, nutrition and immune defence. *Clin. Sci.* **2000**, *99*, 93–104. [CrossRef]
52. Lacroix, I.M.; Li-Chan, E.C. Evaluation of the potential of dietary proteins as precursors of dipeptidyl peptidase (DPP)-IV inhibitors by an in silico approach. *J. Funct. Foods* **2012**, *4*, 403–422. [CrossRef]
53. Nongonierma, A.B.; FitzGerald, R.J. An in silico model to predict the potential of dietary proteins as sources of dipeptidyl peptidase IV (DPP-IV) inhibitory peptides. *Food Chem.* **2014**, *165*, 489–498. [CrossRef]
54. Deacon, C.F.; Nauck, M.A.; Toft-Nielsen, M.; Pridal, L.; Willms, B.; Holst, J.J. Both subcutaneously and intravenously administered glucagon-like peptide I are rapidly degraded from the NH2-terminus in type II diabetic patients and in healthy subjects. *Diabetes* **1995**, *44*, 1126–1131. [CrossRef]
55. Reimann, F.; Williams, L.; da Silva Xavier, G.; Rutter, G.A.; Gribble, F.M. Glutamine potently stimulates glucagon-like peptide-1 secretion from GLUTag cells. *Diabetologia* **2004**, *47*, 1592–1601. [CrossRef] [PubMed]
56. Reimann, F.; Ward, P.S.; Gribble, F.M. Signaling mechanisms underlying the release of glucagon-like peptide 1. *Diabetes* **2006**, *55*, S78–S85. [CrossRef]
57. Reimer, R.A. Meat hydrolysate and essential amino acid-induced glucagon-like peptide-1 secretion, in the human NCI-H716 enteroendocrine cell line, is regulated by extracellular signal-regulated kinase1/2 and p38 mitogen-activated protein kinases. *J. Endocrinol.* **2006**, *191*, 159–170. [CrossRef] [PubMed]
58. Gameiro, A.; Reimann, F.; Habib, A.M.; O'malley, D.; Williams, L.; Simpson, A.K.; Gribble, F.M. The neurotransmitters glycine and GABA stimulate glucagon-like peptide-1 release from the GLUTag cell line. *J. Physiol.* **2005**, *569*, 761–772. [CrossRef]

59. Tolhurst, G.; Zheng, Y.; Parker, H.E.; Habib, A.M.; Reimann, F.; Gribble, F.M. Glutamine triggers and potentiates glucagon-like peptide-1 secretion by raising cytosolic Ca2+ and cAMP. *Endocrinology* **2011**, *152*, 405–413. [CrossRef]
60. Diakogiannaki, E.; Pais, R.; Tolhurst, G.; Parker, H.E.; Horscroft, J.; Rauscher, B.; Zietek, T.; Daniel, H.; Gribble, F.M.; Reimann, F. Oligopeptides stimulate glucagon-like peptide-1 secretion in mice through proton-coupled uptake and the calcium-sensing receptor. *Diabetologia* **2013**, *56*, 2688–2696. [CrossRef]

MDPI
St. Alban-Anlage 66
4052 Basel
Switzerland
Tel. +41 61 683 77 34
Fax +41 61 302 89 18
www.mdpi.com

Foods Editorial Office
E-mail: foods@mdpi.com
www.mdpi.com/journal/foods

www.ingramcontent.com/pod-product-compliance
Ingram Content Group UK Ltd.
Pitfield, Milton Keynes, MK11 3LW, UK
UKHW060124010826
14089UKWH00006B/148